DON'T FORGET THE HUMANS

Sustainable Progress in the Age of AI

Joshua Gideon

HOUSE OF GIDEON

PUBLISHING

ISBN-13: 979-8-9991146-0-0 Paperback
ISBN-13: 979-8-9991146-1-7 Digital online
ISBN-13: 979-8-9991146-2-4 Hardback

Cover design by: Olivia McCauley

Printed in the United States of America

Contents

Legal Disclaimer

AI Disclaimer

In writing this book about preserving human dignity alongside technological advancement, I faced an ironic challenge: How much should I use AI tools in the process? After reflection, I chose a deliberate balance.

What I Used AI For

- Initial research compilation
- Structure refinement
- Editing suggestions
- Idea synthesis

What Remained Purely Human

- The initial draft in all its messy glory
- Personal experiences and reflections
- Emotional insights
- Ethical judgment calls
- The many fine humans who advised on editing, structure, readability, and so many more things, I could never list them all here.
- Final decision-making on all content

My Commitment

Every word that made it into this final book passed through human judgment. Every single one. Every story came from lived experience. Every insight emerged from genuine reflection. AI helped me organize and refine, but the heart, the soul, and the responsibility for this message remain entirely human.

Living the Message

In this process, I lived the central message of the book: Technology serves us best when we remain conscious, intentional, and actively engaged with its outputs rather than passive recipients of its suggestions.

I didn't forget the human while writing this book, and I used the process as a means of practicing the principles I advocate.

Joshua A. Gideon

This book is dedicated to:

To my daughter Allison —

Your voice echoed beside mine through every chapter of this book.

You were my sounding board, my compass, and my constant reminder of why this book matters.

Thank you for boldly listening, questioning, and believing in this book even when I wavered. I love you so much and I am so proud of the woman you have become.

To Grant Cunningham —

Thank you for your friendship and mentorship. Your friendship has made me a better person.

You've shown me how to write with integrity and live with purpose.

Thank you for being the example I often fall short of but always reach toward.

To Mr. Larry McClellan —

You saw a flicker of potential in a struggling kid and fanned it into flame.

Thank you for teaching me that curiosity can be a lifeline, and for believing I was worth the effort.

And to every human this book is for —

May you be seen, heard, and never forgotten.

Acknowledgments

This book would not exist without the many hands, hearts, and brains that helped shape it, both directly and indirectly. If these pages feel human, it's because of the humans behind them.

To my editor, Kathy Allard — thank you for walking alongside me through the complexities and momentum of non-fiction writing. Your polish made the book shine without losing its message. Your thought-provoking notes and suggestions helped me more than you will ever know.

To my coworkers — your encouragement during hallway chats and late-night messages kept me going. Thank you for your encouragement and support throughout this journey.

To my early readers and reviewers — thank you for your generous feedback, your eye for what needed to be clearer, and your honest reactions to what landed (and what didn't). You helped make this book stronger and more accessible.

Finally, to the reader — thank you for choosing this book in a world full of noise. I don't take your attention for granted. I hope you walk away not just informed, but seen.

Part One

The Problem

Introduction
We All Start Somewhere

As a cybersecurity professional, I hear a wide range of questions every week. People are scared by, intrigued by, and puzzled by AI, and a lot of misinformation is floating around. In this book, I draw on my nearly 30 years as a cybersecurity leader to demystify AI, clarify what it can and cannot do, and give my advice for those navigating an uncertain future.

But first, how did we get here, or more specifically, how did I get here? Although this book is as packed with information about AI as I have just promised, it is also partly a memoir. I mix some autobiography, anecdotes, and life lessons into the following pages. My intention is not just to lighten the at-times heavy topic of AI, but also to reinforce my points about technology and our rapidly changing world, and to give the reader – all readers – some reference points, some "Yeah, I know all about that!" or "I did that too!" moments.

So … how did I get here? How did I get to the point that I've been repeatedly asked to write a book sharing my knowledge of AI? The answer might surprise you, because it begins with defying the odds.

"I shouldn't be here."

Those words echo in my head as I drive home from the office on a sunny Indiana day. I sometimes put the top down on my Jeep so I can decompress on the way home. The hum of the engine, the feel of air flowing through my hair, and the fresh smell of tilled soil wafting through the country air along the back roads of Indiana accompany me as I leisurely drive home. I'm a cybersecurity leader now, something that still surprises me when I think about it or hear others mention it. Cybersecurity is a very technical responsibility that scratches the technical itch I developed many years ago. Beyond the glow of my screen, I carry a burden every day: the responsibility to remember the human beings behind the technology.

Because I've seen what happens when we forget them.

I was born almost a month past my due date. My parents were not financially well-off at the time and were relying on the generosity of others for help. My mother's doctor went on vacation around the time she was due to give birth to me. My parents think the doctor thought my mom would go into labor while he was on vacation, and the worst-case scenario would be that someone else delivered me. Besides, they would probably be writing off much of that bill anyway. However, I was not born when they expected, and the waiting came with a price.

My mother endured weeks of preeclampsia waiting for her doctor to return from vacation; her feet were so swollen that the skin cracked, her body breaking at the seams. She was balancing fear with not being a burden on anyone. When I finally arrived, I was a very sick baby. Due to bronchitis, pneumonia, and severe asthma attacks, I was a frequent visitor to hospital ERs even before I could walk. My lungs were not in good shape. My earliest memories aren't of playgrounds or holidays, but of plastic hospital bracelets on my arm, breathing treatments, and the gentle sound of a nebulizer in a darkened room. That sound still puts me to sleep to this day.

By the time I reached third grade, my life was already a patchwork of hospital visits, missed school days, and frustrated teachers. I missed so much school that I had to repeat the grade. It was devastating. It wasn't just the illnesses that kept me home; it was also because people had already decided who I was. I remember one moment that particularly stung: My first third-grade teacher in a parent-teacher conference said to my mother, "He's just lazy. If he doesn't stop being lazy, he'll never amount to anything." She didn't whisper it, didn't soften it. She said it right in front of me, as if I weren't even in the room.

Those words cut deep. They followed me into the placement testing at my new school the following year. However, something surprising happened: They discovered that I wasn't slow or lazy. My test scores painted a much different picture. The kind older man conducting the test placed his reading glasses on me, and things changed. In the end, they discovered I was severely nearsighted. When I finally got my first pair of glasses, the world around me transformed. I remember noticing the details of leaves on trees, as well as the details of individual letters in a sentence, and I was able to pick up on facial expressions that had previously been foreign to me. I could see the chalkboard, and reading became so much easier.

Yet even with glasses and a fresh start, the old insecurities didn't disappear overnight.

Those early years were tough: I was the statistical failure, the kid who should have dropped out, the one labeled as "never amounting to anything." On paper, I was destined to become another statistic of the public education system. Yet here I am, decades later, carrying the responsibility of protecting systems that millions depend on. The fragility that taught me to pay attention, the early failures that forced me to develop resilience, and the outsider status that sharpened my observation weren't obstacles to overcome. They became the foundation of the leader I would become.

The Spark of Technology

If fragility shaped my awareness, curiosity sparked my spirit.

It started with a large brown box in the mail. When I was nine years old, my father purchased an IBM 5160 XT/286 computer kit. When that box arrived on our doorstep, I didn't just see a machine; I saw a doorway into another world. My curiosity took a significant leap forward that day. I was still recovering from years of illness, and suddenly holding the key to a universe I'd only glimpsed in dreams. I tore into the manuals, devouring them like maps that might lead to some buried treasure. By the time my dad and his friend were ready to set up the machine, I had memorized the books, and the diagrams were etched into my mind. As they fumbled with cords and DOS prompts, I piped up nervously, offering advice. My heart raced as I showed them how to use this new technology. For weeks, my dad nervously watched over my shoulder as we set up the new software and eventually wrote menus in batch files (for those of you who remember that). I didn't just help my dad configure a computer; I realized I could shape the world around me. I became, as I often joke, the family "computer guy" long before any of us understood where that would lead me.

But here's the thing: Even as I fell in love with technology, I never became its blind devotee. Maybe it was all those early brushes with mortality, or maybe it was just my nature, but even then, I had a skeptic's heart beating alongside my wonder. I wasn't a Luddite, not by a long shot. I adopted new tools, acquired new knowledge, and pursued innovation wherever it arose. However, I've always viewed new technology through a hacker's lens, a lens of risk, curiosity, and balance. I wanted to understand how things worked, but also where they might break, where they might harm, and where they might fail the humans they were supposed to serve.

That curiosity opened a door for me and got me noticed. I was fortunate enough to get into Mr. Larry D. McClellan's class. We all knew him as Mr. M. This superhuman of a man unleashed the world of science and curiosity in me. He was able to challenge me in a way that no one had ever done before. It was like a switch flipped inside me. It was just what I needed to recover from my rock bottom. He kept me from recess many times to do my work. He found my strengths and encouraged them. He broke through my fears without breaking my will. I remember the day he let me loose in the library. I could check out any book I wanted. This sparked my desire to learn even more.

I was still not the best student. My mom had to come into school every night to go through my desk and find the homework I had stuffed in there, which I had forgotten about because I was off on some rabbit hole learning about a new thing. Mr. McClellan opened my world to science, experimenting, and questioning. I am forever grateful to this wonderful man who saved me from statistical failure by encouraging me to stay curious and try the impossible.

That mindset has followed me into every chapter of my life, and it's never been more relevant than now, in the age of AI. I'm not against artificial intelligence. I'm not running from it, and I'm certainly not dismissing its potential. But I'm not willing to ride the hype train, either. My compass has always been: *Does this serve people? Does it protect their dignity, their autonomy, their meaning? Or does it sacrifice them on the altar of progress?* That's the difference between responsible innovation and reckless acceleration. That's the posture I want to model, and the lens I hope to share with you in this book.

Why This Book Matters to Me

Throughout my years in the technology industry, I've observed several patterns that seem to repeat. These observations are what have driven me to write this book. It's not academic curiosity or professional ego; it's something deeper and more urgent.

I've spent nearly 30 years professionally in the technology industry, and several more before I was paid — helping family and teachers fix computers, troubleshooting hardware in computer repair shops, and serving as a cybersecurity leader. From those early personal computers to cloud computing, from basic encryption to AI-powered defense systems, I've witnessed decades of transformation. Yes, I was around before the internet. (Fun fact: when I was in junior high school in the 1990s, I ran a dial-up bulletin board system devoted to hacker culture.) My journey has taught me something that doesn't get discussed enough in tech circles: The most critical failures aren't technical. They're human.

Early in my cybersecurity career, I learned a painful lesson about the consequences of ego. I was in my 20s, a newly minted junior security engineer and eager to prove myself. I had finally arrived at the table, or so I thought, trusted to manage firewalls, monitor network traffic, and sniff out potential threats. It was my first taste of real authority, and I wore it like a badge of honor. To me, this was proof that I'd finally outrun the shadow of that "lazy" kid from third grade.

One day, while reviewing content filter logs, I came across something that caught my attention: an employee browsing an "inappropriate" website during office hours. Jackpot. I jumped into action, pulling IP addresses, login records, and timestamps. I created a polished report from the content filtering system that clearly showed a flag in inappropriate usage. I marched it straight to Personnel Management (the precursor to Human Resources), eager to demonstrate my value and show that I was worthy of the trust placed in me.

The employee in question was a quiet and kind woman in the office. I didn't know her well, but I felt a jolt of vindication as she was called in, even if I was a little shocked. As we sat in that room, I laid out my evidence, ready to show how vigilant and competent I was. But as she looked at the papers in front of her, her face turned pale. And then she spoke, her voice barely above a whisper.

She asked if I had looked at the website. I had not; I was solely going off the data the system had given me. She then told us she had been diagnosed with breast cancer a few months earlier. After a mastectomy that likely saved her life, she was quietly researching reconstructive surgery options, medical sites that, because of graphic images, had triggered the content filter. She hadn't wanted anyone at work to know.

My investigation had exposed something deeply personal, something she had fought to keep private.

In that moment, the air left the room. My excitement curdled into horror. My triumph collapsed into shame. I saw myself through her eyes, not as a protector but as an intruder who had violated her dignity in the name of security.

That day, I learned the most painful lesson of my career: Behind every log, every alert, every policy violation, is a human being with a story I can't see. I learned that vigilance without empathy is a form of cruelty. That competence without compassion is dangerous. That the power we wield as security professionals can wound as easily as it can protect.

Driving home that evening, I replayed the moment over and over. I saw her face, her quiet dignity in the face of my arrogance, her grace in accepting my apology. I realized that I had let ego lead me, that I had been more invested in proving my worth than in protecting hers. I had forgotten the human.

In the weeks, months, and years that followed, I made a conscious effort to ensure I didn't forget about the people in everything I did. I stopped making jokes about the PEBKAC (Problem Exists Between Keyboard and Chair) or the ID10T errors. I became a more empathetic helper. I attempted to promote prioritizing outreach over punishment. I pushed to create channels for understanding before action. I made a silent promise to myself: Never again would I let the thrill of catching a problem blind me to the person inside it.

This experience fundamentally shaped how I approach technology leadership. It taught me that the most sophisticated systems are worthless if they forget the humans they're meant to serve. As I progressed from hands-on technical roles to leadership positions, I carried this lesson with me, always asking: "Who might be hurt by this? Who are we not seeing? What human needs are we overlooking?"

A Promise to the Reader

I'm not writing this book as someone who has all the answers about the complex intersection of AI, technology, ethics, and humanity. In fact, much of what I am writing is likely nothing new. I am not the first to write about these topics, and I hope I am not the last. I'm writing this as someone who has made mistakes, learned from them, and discovered something I think is profound. The future of technology isn't predetermined. It's being shaped right now by the choices we humans make, the values we prioritize, and the courage we bring to difficult conversations.

Whether you're a curious outsider wondering how to navigate an increasingly automated world, a rising leader grappling with ethical dilemmas, or someone wrestling with doubt about your place in this technological revolution, this book is for you. It's for anyone who has ever felt the gnawing sensation that something important is being lost in our rush toward efficiency.

I promise you three things:

First, I won't offer you fear-mongering. While I'll honestly examine the risks we face, my goal isn't to scare you into hiding. Fear without hope is not helpful to anyone. Instead, I want to help you see these challenges clearly so we can address them rationally and thoughtfully.

Second, I won't give you shallow optimism. I don't believe that technology is inherently good or evil. It's a tool, like any other, that assists humans. Too often, we're sold the myth that innovation alone will solve our deepest problems. That's not only untrue, but also dangerous. Real solutions, solutions only the best of humanity can create, require us to wrestle with the complex intersections of technology, ethics, and human nature.

Third, I will offer you an honest, hopeful guide to building a humane, resilient future alongside technology. We don't have to choose between progress and humanity. We can build systems that enhance rather than diminish our essential nature. But this requires intentionality, wisdom, and the courage to put human well-being at the center of technological development.

Throughout this book, I'll share a few more stories from my journey, not as universal prescriptions, but as invitations to reflect on your relationship with technology. I'll introduce you to frameworks for thinking about technological change that honor both innovation and human dignity. Most importantly, I'll challenge you to see yourself not as a passive recipient of technological change, but as an active participant in shaping its direction. And let me be clear: that invitation isn't just for tech leaders or decision makers. This book is for anyone whose life is shaped by technology. That means all of us. You don't need to control the systems to have a voice in how they affect your humanity.

What gives me hope is that we're having these conversations now. You're reading these words, questioning assumptions, and seeking understanding. That's where change begins, not in boardrooms or laboratories, but in individual human minds and hearts that refuse to accept dehumanization as inevitable.

The technological revolution is here. AI will continue to advance, and automation will continue to expand. Robots will be developed that can perform tasks typically performed by humans. Despite that, we still have a choice about what kind of future we build with these tools. Will we create systems that serve humans, or will we create a world where humans exist merely to serve the systems?

I believe we can choose wisely. I believe we can build better. I believe we can remember the humans.

I believe this not because I'm naive about the challenges ahead, but because I've seen what happens when we forget, and I've also seen the extraordinary resilience of the human spirit when we remember what truly matters.

This book is my contribution to that remembrance. It's an invitation to join a crucial conversation about our shared technological future. It's my way of honoring all the people who saw potential in a sickly, struggling kid who "wasn't supposed to amount to anything."

Introduction

They remembered the human in me when I couldn't see it myself. Now it's our turn to remember the humans in the systems we're building.

Chapter 1
The Machines Are Winning

How the Machines Are Winning

The shift happened while we weren't looking.

We've designed technology that no longer assists, it decides. Algorithms don't just recommend, they manipulate and define reality. Automation doesn't simply replace tasks; it reshapes purpose itself.

This recognition resonates differently depending on one's perspective. For some, it's liberation from mundane work. For others, an existential threat. For most, it's a quiet, uneasy feeling that something fundamental is changing in our human world.

The psychological impact extends beyond job displacement. We're witnessing machines that challenge core assumptions about human uniqueness. When AI generates art, composes music, and solves problems with superhuman accuracy, the questions become unavoidable: Is AI better than we are? Is it too late to stop AI from taking over? What remains, and should remain, distinctly human?

This isn't fearmongering. It's confronting reality. The trajectory is clear:

Systems that once augmented human capability now sometimes surpass it.

The Rise of Automation Across Industries

The automation wave has expanded beyond manufacturing floors to encompass every sector of human work. Legal firms deploy AI to review contracts faster than an army of attorneys. Healthcare systems use algorithms to diagnose conditions, sometimes with greater accuracy than experienced physicians. Financial institutions have replaced human traders with systems that execute millions of transactions in microseconds.

In journalism, AI writes basic news reports. In education, automated systems grade essays and provide personalized learning paths. Even creative industries, once considered safe havens of human uniqueness, now face AI that generates music, writes code, and designs graphics indistinguishable from human work.

The speed of this transformation has exceeded anything in human history. An entire industry can be revolutionized in months rather than decades. The psychological impact of that reality on workers is immeasurable — the slow-burning anxiety that comes from watching your expertise become algorithmic and your judgment become replaceable.

The Creeping Influence of AI into Daily Life

We carry this technology in our pockets, sleep with it on our nightstands, and wake up to its recommendations. Our smartphones predict our next words before we type them. Navigation systems reroute our drives based on real-time traffic analysis. Streaming services shape our entertainment choices, often determining what we enjoy before we know it ourselves.

Social media algorithms curate reality itself, creating feedback loops that simultaneously connect and isolate us. We scroll through feeds designed to maximize engagement rather than meaningful connection. The line between conscious choice and algorithmic manipulation grows more blurred every day.

Chapter 1

Personal assistants schedule our calendars, smart homes adjust our environments, and recommendation engines influence everything from our shopping to our reading lists. We've invited systems into intimate spaces and have begun to trust them with decisions that subtly reshape our daily routines and have a profound impact on our choices.

Early Warning Signs We Ignored

To borrow a term from the intelligence world, the "signals" appeared years ago, but we rationalized them away. When factories automated, we told workers to "up-skill." When algorithms started managing workforces, we celebrated optimization. When AI began diagnosing medical conditions, we praised efficiency over the erosion of doctor-patient relationships. We did all this while ignoring the human cost.

After accepting what seemed like a dream opportunity to lead a team and build a managed security services division at a Managed Service Provider (MSP), I witnessed this firsthand. Instead of my dream job, I found myself in a dehumanizing environment where the narcissistic owner demanded total control. He forced employees to work 60 hours weekly for a salary while double-billing customers. Personal vehicles had to meet his exacting standards — always washed, full of gas, and spotless inside. Numbers measured everything and everyone. How many hours were you billing? Did you drive the shortest route to the customer to reduce your mileage reimbursement? What did you upsell to the customers this week? Were you able to bill one customer while working on something for another? The metrics changed faster than you could keep up with them.

This dehumanizing environment perfectly exemplifies systems designed around machine logic rather than human needs. Employees weren't valued as people with expertise and judgment, but as inter-changeable resources to be optimized and controlled. The metrics mattered more than the humans generating them. The money mattered more than the well-being of the people who were taking care of their clients. Inevitably, customers began to notice. Service quality suffered as burnt-out technicians struggled to maintain the facade of enthusiasm while being crushed by impossible demands. The irony was painful yet predictable. It was a system designed solely for profit, a system that ultimately undermined the very customer relationships that generated the profit in the first place.

Eventually, the burnout became unbearable. I escaped after reaching out to a retired police officer friend who had been doing executive protection work. He connected me with a local company, and I burned through all my vacation time and savings to take an executive protection certification class. I secured some contract jobs and submitted my resignation to the company. It was one of the best decisions I ever made.

My experience reflects a broader societal pattern. We've ignored the rising suicide rates in communities hollowed out by automation. We've dismissed the growing anxiety among professionals whose expertise was being commodified. We've overlooked the quiet desperation of workers who felt themselves becoming invisible within systems designed around machine logic.

These warning signs weren't dramatic but cumulative, a slow increase in burnout rates, growing instances of "quiet quitting," and rising levels of medication for anxiety and depression. These weren't isolated incidents, but symptoms of a deeper transformation, one so vast that it is systematically redefining human worth in terms of productivity and compliance rather than creativity and judgment.

The Historical Mirror

Previous technological revolutions offer sobering lessons, but this time the parallels are imperfect.

The Industrial Revolution of the 18th and 19th centuries displaced craftsmen over decades. Communities adapted, however painfully, across generations. Today's transformation compresses similar disruption into months or years. Machine learning evolves daily, automation scales weekly. Human psychological adaptation, which requires time for processing and cultural adjustment, hasn't been able to keep pace.

Each historical transition promised to elevate humanity, but in reality, it delivered unintended consequences and unforeseen costs. The printing press democratized knowledge but also enabled propaganda. Radio unified national audiences but also became a tool for mass manipulation. The internet promised connection yet delivered isolation for many.

The pattern persists: initial euphoria, rapid adoption, unintended consequences, social upheaval. Yet each time we tell ourselves we'll manage the transition better. We'll protect human values. We'll maintain control.

History suggests otherwise. The real lesson isn't that technology is inherently dangerous. It's that we consistently underestimate its power to reshape human experience.

Echoes of Past Industrial Revolutions

The first Industrial Revolution mechanized production, transforming agricultural societies into manufacturing powerhouses. Skilled artisans watched centuries-old crafts become obsolete. Weavers, blacksmiths, and other craftsmen, whose identities were intertwined with their work, faced displacement not just economically but in ways that shattered their sense of meaning.

The second wave brought assembly lines, breaking complex work into repetitive tasks. Frederick Taylor's "scientific management" reduced human workers to components in an industrial machine. Pride in craftsmanship gave way to the monotony of repetitive work. Workers became interchangeable parts in a system optimized for output over human pride and craftsmanship.

The digital revolution promised to liberate us from drudgery yet created new forms of it. Email chained us to constant connectivity. Social media has addicted us to dopamine feedback loops. The internet, which was supposed to democratize information, instead paved the way for echo chambers and misinformation on an unprecedented scale.

The Human Cost of "Progress" Then

History books sometimes celebrate innovation while frequently glossing over human suffering and carnage. The Industrial Revolution's progress came at the cost of child labor, dangerous working conditions, and the destruction of traditional communities. Families were separated as work moved from homes to factories. Generational knowledge was lost as craft gave way to mass production.

The Gilded Age of the late 19th century saw unprecedented wealth creation alongside extreme poverty. The prosperity of industrialists was built on the exploitation of immigrants, women, and children working in unsafe conditions. Progress was measured in terms of output and profit rather than in human welfare.

Even the digital revolution, initially hailed as a democratizing force, has created a new class system. Tech giants accumulate wealth while gig workers struggle without benefits or job security. The promise of remote work has often meant the dissolution of work-life boundaries, leaving people in a state of perpetual availability.

What We Failed To Carry Forward

With each technological transition, we promised ourselves we'd learned from the mistakes of the previous one. This time we'd prioritize human welfare. We'd ensure that progress served people, not shareholders. We'd protect workers' dignity and the unity of communities.

Yet, with each revolution, we have forgotten these promises. The lessons of industrial exploitation were ignored as we built digital sweatshops. The hard-won labor protections of the 20th century eroded in the face of automation and platform work. The social safety nets constructed after the Great Depression can be argued to have been a short-term success and a long-term failure. They have been systematically dismantled in the name of efficiency.

We've failed to carry forward the most crucial lesson: Technological progress without ethical consideration creates social division, human suffering, and ultimately, systemic instability. Innovation divorced from wisdom doesn't serve humanity; it consumes it.

The Ethical Reckoning

Our technological capability has dramatically outpaced our ethical frameworks.

Consider current realities: AI systems make hiring decisions with biases we can't detect. Algorithms determine creditworthiness using opaque criteria that perpetuate inequality. Automated content moderation lacks a nuanced understanding of context and intent.

These aren't hypothetical concerns. They're operating at scale, affecting millions of lives with minimal oversight.

As systems grow more complex, human oversight becomes practically impossible. Machine learning models process millions of data points to reach conclusions no human could verify. We've created decision-making systems we can't fully understand, yet we trust them with choices that profoundly impact human lives.

Even greater danger emerges when efficiency becomes conflated with ethics. When speed and scale dominate success metrics, human considerations become "friction" to be eliminated. Dignity becomes optional. Meaning becomes secondary to optimization.

This trajectory isn't sustainable. Systems that prioritize pure efficiency over human well-being create conditions for their rejection, whether through quiet withdrawal, active resistance, or systemic failure.

Where Technology Outpaces Conscience

We've built systems that operate at a speed beyond the limits of ethical debate. High-frequency trading algorithms execute millions of transactions per second, generating market volatility that exceeds human comprehension. Social media algorithms amplify outrage faster than we can develop emotional resilience. Surveillance systems collect data at rates that overwhelm our capacity for meaningful privacy protection.

This acceleration of technological capability has created an "ethics lag," a growing gap between what we can do and what we should do. By the time we recognize the ethical implications of a new technology, it's often already embedded in the lives of millions and defended by powerful economic interests. I haven't even mentioned how difficult it is to regulate politically. We are forced to accept and adapt, seemingly with no choice.

This lag isn't just about regulation. It's about the fundamental pace of moral reasoning in human communities. Ethical frameworks develop through dialogue, experience, and cultural evolution, processes that require time for reflection and consensus building. Technology, particularly AI, evolves exponentially, outpacing our society's ability to comprehend its implications.

The result is technological orphans, systems deployed without ethical parents, operating in moral vacuums where optimization replaces wisdom, and market forces substitute for values.

Systems Designed for Profit vs. People

The architecture of most technological systems reveals their true priorities. Social media platforms prioritize engagement over well-being. E-commerce sites optimize for purchases, not informed decisions. Dating apps prioritize usage time over meaningful connections. Gig economy platforms extract maximum value from workers while minimizing benefits and protections.

These aren't accidental outcomes but deliberate design choices. When venture capital demands exponential growth, when stock markets reward quarterly profit over long-term sustainability, and when success metrics prioritize scale over impact, human welfare becomes a secondary concern at best.

The profit motive, which has driven much valuable innovation, becomes distorted when unchecked by ethical considerations. Silicon Valley's reputation for moving fast and breaking things assumes that human systems are as replaceable as code. Yet broken trust, damaged communities, and traumatized individuals can't be fixed with software updates.

This design paradigm creates what economists call "negative externalities," costs imposed on society that don't appear on corporate balance sheets. When a social media algorithm contributes to teen depression, the human costs are real even if they're not monetized.

The Danger of Forgetting the Human

Perhaps the most significant ethical risk is the gradual abstraction of human beings into data points. When we interact with people through interfaces, make decisions based on aggregated statistics, and optimize for metrics rather than meaning, the individual human experience becomes invisible.

This abstraction enables decisions that would be unconscionable in person. The danger compounds when abstraction meets automation. Systems designed without human oversight can perpetuate and amplify the biases we're trying to escape. An AI trained on historical hiring data may encode decades of discrimination. An algorithm designed to minimize costs may recommend denying healthcare to those who need it most.

Forgetting the human enables a form of structural violence — harm built into systems rather than emerging from individual malice. It creates a world where suffering becomes a byproduct of efficiency, where dignity becomes an obstacle to optimization, and where humans are sacrificed to the gods of algorithmic logic.

The Human Horizon

Despite these challenges, specific human capabilities remain irreplaceable.

Machines cannot replicate the capacity for moral reasoning in complex and unpredictable situations. They cannot navigate ambiguity or understand the concept of contradictions. Emotional intelligence emerges only from lived experience. Living creatures are the only beings capable of true creativity and intuition.

These aren't romantic notions about "human superiority." They're practical observations about what remains essential as technology advances.

The path forward requires intentionality rather than resistance. We need leaders who understand both technological capability and human needs. We need systems designed for humans, not merely mindless efficiency. We need sustained dialogue between technical innovation and ethical consideration.

This is a shift in how we approach technological development. Beyond asking "Can we build this?" we must consistently ask "Should we build this?" and "Whom does this truly serve?"

The machines appear to be winning because we've let them play a game designed for their strengths while neglecting our own. The challenge isn't competing with machines but ensuring that human values shape how we deploy them.

Why Creativity, Empathy, and Judgment Matter Now

As AI excels at pattern recognition and data processing, uniquely human qualities become increasingly valuable, not less so. Creativity involves more than recombining existing elements, and it requires the intuitive leaps that emerge from lived experience and emotional depth. AI can generate variations within parameters, but genuine innovation often comes from breaking those parameters entirely.

Empathy remains strictly biological. While AI can simulate compassionate responses, it cannot feel the visceral understanding that comes from sharing human vulnerability. Think about the nurse who recognizes subtle signs of distress, the teacher who senses when a student is struggling beyond academics, and the manager who knows when someone needs support rather than pressure. These human capacities create trust and healing that no algorithm can replicate.

Judgment, particularly ethical judgment, requires exactly the tolerance for ambiguity that machines lack. Human decision-making incorporates context, history, values, and what philosophers call "practical wisdom" — the ability to do the right thing in unique circumstances that have no defined rulebook. This isn't inefficiency; it's the very essence of moral reasoning.

The Invitation to Resist Dehumanization

Resisting dehumanization isn't about rejecting technology; it's about insisting it serves human ends. Resistance begins with small, personal choices: prioritizing face-to-face conversation over texting, choosing human customer service over chatbots when possible, and supporting businesses that value workers over algorithms.

On a systemic level, resistance means demanding transparency in automated decisions, advocating for human oversight in consequential systems, and insisting that efficiency serve rather than replace human judgment. It means designing workplaces that amplify human capability rather than reducing humans to components.

Fundamentally, it means remembering what technology is for. Every system, algorithm, and automated process should ultimately serve the goal of helping humanity. When we lose sight of this purpose and optimize for metrics that fail to capture human values, we create the conditions for our obsolescence.

A Call to Leadership and Stewardship

The technological moment demands new forms of leadership: people who understand both code and conscience, who can navigate between innovation and ethics, and who prioritize long-term human well-being over short-term metric wins.

This leadership isn't limited to tech executives or policymakers. Every person working with or around technology has opportunities for ethical stewardship: the developer who insists on inclusive design, the manager who protects human judgment in automated workflows, and the individual who models healthy boundaries with technology, to name just three examples.

We need leaders who ask different questions: not just "How fast can we scale?" but "How well can we serve our customers?" Not just "What can we automate?" but "What should we preserve?" Not just "How do we optimize?" but "What values do we embody?"

More importantly, we need collective courage to prioritize human dignity over technological efficiency when they conflict. This requires recognizing that some things shouldn't be optimized, some decisions shouldn't be automated, and machines shouldn't mediate some experiences.

The Path Forward

Technology isn't our enemy, nor is it our savior. It's a powerful amplifier of human intention.

The most critical technological decision isn't which capabilities to develop, but which values to prioritize as we create them. This requires courage to slow down when the culture demands acceleration, wisdom to ask hard questions when simple solutions seem available, and commitment to preserving human agency when automated efficiency beckons.

I intend to explore these tensions not as abstract philosophy, but as urgent and practical concerns. I have attempted to ensure that each chapter examines different facets of technological disruption and human adaptation, always returning to the central question: How do we build systems and technology that enhance, rather than diminish, human dignity?

The future isn't predetermined. It's being shaped by choices we make daily, in product design, policy decisions, personal boundaries, and social expectations. Every technological choice carries moral weight, whether acknowledged or not.

We stand at a crossroads, where the decisions made over the next decade will echo for generations. The choices we make now about AI governance, worker protections, data privacy, and technological dependency will determine whether we create a world that serves humans or one where humans serve machines.

This isn't about rejecting progress. It's about insisting that progress include all of us, that no one becomes obsolete, and that meaning matters more than metrics.

The Real Challenge

The machines are winning specific battles, in terms of efficiency, speed, pattern recognition, and data processing. But victory itself remains a human concept, defined by human values.

My career in cybersecurity has repeatedly confirmed that the most significant vulnerabilities aren't technical. They're human, caused by the erosion of judgment, loss of context, and failure to see people behind data points.

As we navigate this transformation, our greatest asset isn't our ability to build increasingly sophisticated systems. It's within our capacity to remain fully human while doing so — to bring wisdom to innovation, add empathy to efficiency, and remember that every system ultimately serves people. When we forget those people, the system eventually fails.

This isn't about choosing between progress and humanity. It's about ensuring they advance together, each strengthening the other.

As the pace of technological change accelerates, we must ask if we can lead it with wisdom, ensuring that as our tools become more powerful, our commitment to human well-being grows stronger.

The transformation is already underway. The machines are not inherently winning; we've simply allowed ourselves to play by their rules. The path forward requires us to redefine success itself, ensuring that human values shape technological development rather than being shaped by it. In this redefinition lies our opportunity to create a future where progress truly serves humans, where innovation enhances rather than erodes our humanity, and where technology amplifies the best of who we are instead of reducing us to what machines can measure.

Chapter 2
Industrial Ghosts:
How We've Forgotten Workers Before

The Illusion of Progress

The story of progress tells itself through grandiose eyes and impressive statistics. We measure advancement in productivity gains, efficiency improvements, and technological milestones. Each new wave of automation is celebrated as a liberation: freedom from mundane tasks, release from physical strain, and the triumph of human ingenuity over limitations.

But there's another story hiding beneath the metrics, one we rarely acknowledge in our rush toward optimization.

When mechanical looms replaced skilled weavers in 19th-century England, society marveled at the sudden availability of affordable textiles to the common people. When assembly lines transformed manufacturing in the 1910s, we celebrated how Ford's Model T brought automobile ownership to ordinary Americans. When industrial robots took over welding and assembly in the 1960s, we praised their precision, consistency, and reduced human error.

Each wave was heralded as an advancement, as liberation from tedium, and as the inevitable march toward something better.

Industrial Ghosts:

Yet beneath those celebrations ran an undercurrent of loss, a loss we've never properly acknowledged or mourned.

The men and women who once transformed raw materials into useful objects with their hands, who saw themselves reflected in what they built, found themselves slowly pushed aside. The thread connecting effort and meaning, labor and dignity, began to unravel. What replaced it wasn't just unemployment, but something even more devastating: a sense of purposelessness.

We talk about industrial change through the sterile language of economics: market transitions, productivity gains, and labor displacement. But the true story isn't in the GDP figures or efficiency charts. It's written in abandoned towns, in fractured families, in the quiet resignation of people who once stood at the center of productive life and then found themselves watching from the periphery.

Progress rarely announces itself as a mistake while it's happening. We celebrate the milestones without seeing the gravestones.

We've forgotten something essential about human nature in our rush toward efficiency. We've forgotten that people need more than wages. They need to matter.

And when meaning is stripped from work, it doesn't simply vanish. It transforms into resentment, despair, and eventually resistance.

Now, here we are again, standing at the edge of another transformation. Companies are building AI, automation, and algorithms that autonomously decide who works, how, and when — with no human intervention. We aren't just repeating history; we're accelerating and amplifying it.

The ghosts of the last Industrial Revolution are whispering warnings. If we fail to hear them and continue treating humans as just another resource to optimize, we won't merely repeat past mistakes.

We'll create wounds too deep for our social fabric to bear.

Chapter 2

The First Age of Automation

I've always been fascinated by archaeology. Movies like *Raiders of the Lost Ark* and *The Mummy* are some of my favorites. I am unapologetic about inserting a dad-joke-level analogical sentence to describe what has happened in the past. Brace yourselves for my tongue-in-cheek humor.

To understand our precarious present state, we must first excavate the past: not just uncovering what happened, but also what we failed to brush away to learn from it. (See, it wasn't that bad.)

Seriously, though, I believe that history speaks to us in patterns, not just facts. And right now, its voice carries a warning. To understand what's coming, we must first look at where we've been.

History is not a story of smooth, uninterrupted progress. It is peppered with stories of domination and defiance, control and rebellion, dehumanization and the struggle to reclaim humanity. And again and again, when powerful systems have tried to crush self-respect, humans have not gone quietly.

Industrial Revolution: The Great Displacement

It began with the rhythmic clacking of power looms in the textile mills of northern England. These machines could produce in hours what had once taken days by hand. Steam engines never tired, never complained, and never demanded better conditions.

To economic historians, this transformation represented progress incarnate: Productivity soared, goods became affordable, and new industries bloomed where none had existed. The industrial age had arrived, promising prosperity and advancement.

But walk just a few streets from those triumphant factories, and you'd find a different reality.

Imagine the weaver: a craftsman whose hands knew the subtle tension of thread, whose eye recognized minute imperfections, and whose skill had been cultivated through years of patient apprenticeship. Now imagine them standing before a machine that renders their life's mastery obsolete in an instant.

Imagine the blacksmith, the cooper, and the wheelwright: each watching as their contributions to community life were systematically dismantled, replaced by standardized parts and mechanized processes.

While William Blake wrote of "dark Satanic mills" to describe spiritual oppression by religious institutions, his imagery resonates with how industrial mechanization created its own spiritual dislocation, watching human craft transform into mechanical reproduction and seeing people reduced to extensions of machines rather than creators.

Communities fractured under this weight. Villages were emptied as workers migrated to industrial centers. Families that had lived and worked together for generations were scattered across factories, mines, and mills; their days no longer marked by seasons and natural rhythms, but by the relentless demand of the factory clock.

It was progress, yes. But it came wrapped in loss — loss of livelihoods, identity, and ways of life that had endured for generations.

And then the Luddites emerged.

They were not mindless destroyers of technology, as history books often caricature them. They were protesting the dehumanization of work, the loss of autonomy, skill, and pride. I've frequently thought about how they must have felt, watching their children go hungry while machines took the work that had defined their ancestors for generations.

They smashed the machines not because they feared progress, but because they feared being reduced to disposable labor. Their uprisings sowed the seeds of the labor movements that followed, movements that would reshape the relationship between workers and owners for centuries to come.

Taylorism & Fordism: The Rise of Optimization

By the dawn of the 20th century, industrialization had undergone significant evolution. It wasn't enough to replace craft with machines; now came the drive to optimize every human movement.

An American mechanical engineer named Frederick Taylor could be considered the father of management consultants. In 1911, he published his book, *The Principles of Scientific Management*. The system he presented broke labor into its smallest components to extract maximum efficiency. Workers were timed with stopwatches, and their movements were analyzed, streamlined, and standardized. The goal was no longer craft but output: uniform, predictable, and quantifiable.

Henry Ford perfected this approach with his assembly line. By breaking down the complex construction of automobiles into a series of simple and repeatable tasks that can be performed by interchangeable workers, he revolutionized the manufacturing industry. The Model T rolled off production lines faster and cheaper than any vehicle before it.

But something profound happened in this transition.

A worker who once built an entire carriage now spent his day tightening the same bolt, over and over. A craftsman who had taken pride in his complete creation now saw only a fragment, disconnected from the finished product, from customer satisfaction, and from any sense of whole achievement.

Work became compartmentalized. The monotony of repetition replaced the satisfaction of creation.

This wasn't merely economic efficiency. It was the reorganization of human identity itself. The worker wasn't valued for knowledge, judgment, or creativity, but for conformity, predictability, and compliance. People became, in Taylor's cold parlance, "hands" — not hearts or minds, just functional appendages of the industrial machine.

The alienation that followed wasn't accidental. It was designed.

Globalization: The Hollowing Out

In the late 20th century, a new disruption arrived: globalization.

Industrial Ghosts:

Factories that had been the lifeblood of American communities for generations began closing, as their operations were relocated to countries with cheaper labor, fewer regulations, and tax advantages. Towns built around single industries, such as steel, coal, and automobiles, watched their foundations crumble.

The "company man" who had given 30 years to one employer with the expectation of security and respect was suddenly expendable. Pink slips replaced gold watches. Loyalty became a liability rather than an asset.

I've visited these hollowed-out towns. I've sat in diners where men in their 50s stared into coffee cups, recounting the day their factory closed like others might describe a death in the family. And that's what it was: the death of identity, purpose, and belonging.

The social fabric unraveled with stunning speed. Rural towns that once bustled with activity fell silent. Little League teams lost their sponsors. Churches saw their congregations dwindle. Community centers that once buzzed with shift workers unwinding after long days now stood silent or had been converted into job placement offices.

Under it all grew something darker: a sense of betrayal and abandonment. The social contract that had promised dignity through work was revealed as disposable, just another cost to cut in the pursuit of profit.

Anticolonial Uprisings

Across Africa, Asia, the Americas, and beyond, colonial empires imposed violence, extraction, and cultural erasure for centuries. The scars remain visible today.

And yet: Uprisings ignited from Haiti to India. Languages were preserved underground even when speaking them meant punishment. Songs of resistance traveled across borders, keeping hope alive.

Decolonization movements swept the globe in the 20th century, reminding the world that domination always has an expiration date. The human need for cultural identity and self-determination cannot be permanently suppressed.

Chapter 2

The Pattern We Missed

What connects these waves of industrial transformation? What thread runs from the power looms of Manchester to the outsourced factories of Indiana?

At each stage, we measured success through a dangerously narrow lens: We tracked GDP growth, not community stability. We counted units produced, not lives enriched. We calculated profit margins, not purpose preserved.

We told displaced workers they would adapt, markets would adjust, and new jobs would emerge from the ashes of the old. Sometimes that happened. Often it didn't.

What we failed to recognize is that adaptation is not automatic. It requires educational, social, and psychological infrastructure. When you rupture a person's connection to usefulness and self-worth, you don't just create unemployment; you create despair.

And now, as technology like AI and autonomous robots loom on our horizon, the pattern repeats itself, only faster, with even less time for communities and individuals to adapt. We tell ourselves it's different this time, that the invisible hand of progress will smooth the transition.

But the ghosts of industrial revolutions past are stirring. And they have seen this story before.

What We Failed To Learn

It would be easy to romanticize the pre-industrial past, to imagine some golden age where every worker whistled while crafting meaningful objects, where communities never fractured, and where progress never demanded sacrifice. That would be dishonest, though.

The truth is more complex. Industrial advancement brought genuine benefits: It lifted millions out of poverty, created new opportunities, and freed humanity from certain forms of brutal physical labor.

Industrial Ghosts:

The problem wasn't mechanization itself. The problem was our failure to account for the human consequences, our willingness to sacrifice human purpose on the altar of efficiency.

We Ignored Dignity

When we reduced people to inputs and outputs, we stripped work of its deeper meaning.

A mill worker might not have composed symphonies with her shuttle, but she occupied a recognized place in the fabric of society. The metal-worker might not have designed the machines he maintained, but he knew his skill kept the factory running. Their work connected them to something larger than themselves: a product, a company, and a sense of belonging in a community.

When automation rendered these roles obsolete, we celebrated the productivity gains while ignoring the toll on humanity. We failed to ask: What happens when a person can no longer see their contribution reflected in the world?

The pattern repeated itself across industrial America. Workers who had spent their careers mastering crafts suddenly found themselves obsolete. Their knowledge, accumulated over decades, became irrelevant in a matter of months. Not just employment was lost, but also identity, purpose, and the sense that their life's work had meaning.

Dignity isn't a luxury reserved for the educated, the creative, and the entrepreneurial. It's the fundamental human need to be seen, to contribute, and to belong. When we designed industrial transitions without considering this need, we not only changed economies but also disrupted them. We unintentionally wounded the souls of many people.

We Normalized Disposability

As machines replaced human labor, we didn't just reduce workforce numbers; we transformed how we perceived human value itself.

The machinist with 20 years of experience wasn't valued for their knowledge; they were "redundant overhead" to be eliminated. The assembly-line worker wasn't seen as a repository of institutional memory; she was an expense that automation could erase. The mining town wasn't considered part of the company's responsibility; it was collateral damage in the pursuit of shareholder value.

This disposability didn't stay contained within factory walls. It seeped into our cultural consciousness, reshaping employment itself: Temporary contracts replaced permanent positions, gig work replaced careers, and "human resources" replaced personnel departments. The very language reveals our shift: from people to resources, and from relationships to transactions.

When humans become interchangeable parts, they internalize that message. They stop seeing themselves as bearers of unique value. And a society of people who don't value themselves becomes brittle, vulnerable to both despair and those who would exploit it.

We Undervalued Adaptation

"They'll find new jobs," we said. "They should learn new skills," we insisted. "Creative destruction is necessary for growth," we rationalized.

These statements contain partial truths but betray a profound misunderstanding of human psychology and community dynamics.

We dramatically underestimated the difficulty of mid-life career transitions. We failed to invest in meaningful retraining programs. We offered little psychological support for those facing identity collapse. We provided no pathway for experienced workers to pass their knowledge to the next generation.

Instead, we watched communities sink into depression, both economic and psychological, while critiquing their "failure to adapt" from the safety of thriving cities and growing industries.

The challenge of retraining displaced workers exposed the gap between political rhetoric and human reality. The idea that every unemployed factory worker could simply "learn to code" overlooked the physical toll of decades spent on assembly lines, the psychological burden of starting over in midlife, and the question of how to honor knowledge that had suddenly lost its market value.

It's not that adaptation is impossible. Adaptation indeed requires resources, support, and, most importantly, respect for what came before. We offered too little of each.

The Lesson We Keep Missing

The fundamental lesson of industrial displacement isn't that technology is bad or that change should be resisted. It's when we design systems without considering human value, meaning, a sense of purpose and belonging, that we create wounds that fester for generations.

Automation doesn't have to dehumanize. Technology doesn't have to erode purpose. Progress doesn't have to leave a trail of wreckage in its wake.

But achieving humane transitions requires intention, investment, and a fundamental shift in what we value.

And now, as AI prepares to transform knowledge work the way machinery transformed physical labor, we stand at another crossroads. Will we repeat the same patterns — ignoring dignity, normalizing disposability, and undervaluing adaptation?

Or will we finally learn what history has been trying to teach us?

Here is the pattern that cuts across every era, every culture, and every technological shift:

When human worth is degraded, autonomy stripped away, and people treated as obstacles or afterthoughts, resistance becomes inevitable.

It may not come on a schedule. It may not look as we expect. But it will come.

Our current age of AI, automation, and digital control will be no exception.

Echoes in the Present

The ghosts of the industrial past didn't vanish; they merely shape shifted.

They no longer manifest in belching smokestacks and thundering machinery. Today, they whisper through fiber-optic cables, flicker in the blue light of screens, and hide behind algorithms that decide who is assigned work, who gets hired, and who is valued.

But make no mistake: The patterns remain, accelerating in ways both familiar and frightening.

Past Pattern and Present Parallel

In factories, skilled craftsmanship was replaced by mechanized processes. In today's offices, knowledge work is being increasingly replaced by artificial intelligence.

Factory workers were reduced to units of production on assembly lines. Today's workers are reduced to metrics on productivity dashboards.

Industrial towns collapsed as manufacturing moved overseas. Today, remote work dissolves workplace communities while maintaining the illusion of connection.

The language has changed. The aesthetics have evolved. However, the fundamental dynamic persists, and human connection is rendered obsolete.

Displacement, But Smarter

We once marveled at machines that could weave fabric or stamp metal. Now we marvel at algorithms that can write essays, draft contracts, diagnose illnesses, or create art.

For a moment, it feels magical, just as those early factories once did.

But beneath the wonder, displacement has already begun. Customer service representatives watch their jobs disappear into chatbots. Paralegals, who once handled document review, now see it being handled by AI. Graphic designers compete with image generators that produce images in seconds, whereas it would take the designers hours.

The promises echo those of previous industrial waves: "This will free you for more creative work." "This will make you more productive." "This will handle the boring parts of your job."

For some, these promises may prove true. But for many others, the lived reality feels less like enhancement and more like erasure: watching skills cultivated over years rendered suddenly obsolete, wondering what part of themselves remains valuable.

The displacement isn't just economic. It's existential: questioning not just what we do, but also who we are when what we do is no longer needed.

The Rise of Digital Cogs

The factory worker once punched a time clock, performed repetitive tasks under the watchful eyes of supervisors, and was measured by the units produced.

Today's knowledge worker logs into systems that track keystrokes, monitor active time, and measure output against benchmarks. The surveillance is often invisible and the control more subtle, but no less complete.

Human performance is continuously quantified, broken down into key performance indicators (KPIs), productivity metrics, and engagement scores. We are no longer just employees or colleagues; we are data points to be optimized.

And when people become data points, something essential begins to wither. The immeasurable human qualities of wisdom, empathy, ethical judgment, and creative intuition are devalued precisely because they resist quantification.

I have watched the cultures at these companies collapse into a pile of ashes. No new ideas that can actually get to market are created, and the dehumanized organization becomes an ouroboros serpent that eats the culture into non-existence.

This is one of the hidden costs of optimization: not just the loss of jobs, but also the loss of self within the jobs that remain.

The Fraying of Social Fabric

When industrial towns collapsed, the damage was visible: abandoned buildings, empty storefronts, and dwindling congregations in once-vibrant churches.

Today's unraveling happens behind closed doors, on screens, in the growing isolation of digital work. But it is no less real: Remote workers often go days without meaningful human interaction beyond video calls. Gig workers piece together livelihoods without colleagues, mentors, or a professional community. Social media creates the feeling of connection while also deepening feelings of loneliness and division.

What happens to the informal mentorship that once happened naturally in workplaces? The friendships forged through daily proximity? The collective identity that comes from shared purpose?

These connections aren't luxuries. They're the invisible infrastructure that builds personal and social resilience. When they fray, we become more vulnerable to extremism, despair, and manipulation.

The Acceleration Trap

Here's what makes our current moment uniquely dangerous: the pace.

Each wave of technological disruption now arrives faster than the last, with less time for adaptation. The mechanization of agriculture, for instance, unfolded over generations. The Industrial Revolution transformed manufacturing over the course of several decades. The digital revolution has reshaped knowledge work over several years. AI threatens to compress changes of similar magnitude into a matter of months.

Yet we tell ourselves the same comforting stories: "Markets will adjust." "New jobs will emerge." "Humans will find their place."

Adaptation — true and healthy adaptation — requires time. It requires social support systems, educational infrastructure, and psychological space to reimagine identity. When change outpaces these human necessities, the result isn't a smooth transition. It's a rupture.

We are running toward a cliff, convinced we'll grow wings on the way down.

Why This Moment Matters

We stand at the edge of a transformation that is more profound than any since the first Industrial Revolution – a transformation not just of industries, but of what it means to be human in a world increasingly run by machines.

If we ignore the lessons of the industrial past, we won't just automate jobs, we will automate away meaning. We will hollow out purpose, community, and human connection, leaving a society that functions efficiently while its people suffocate spiritually.

The ghosts of industrial displacement are whispering a warning to us. The question is: Will we listen this time?

Patterns I've Witnessed

Throughout my career in cybersecurity and technology consulting, I've witnessed a recurring pattern across various industries. It's not one dramatic story, but rather a slow, persistent erosion that happens in countless variations.

When new technologies are introduced, the pattern goes like this: First, there's the excitement of innovation. New systems promise efficiency, accuracy, and cost savings. Teams are told this will free them to focus on "high-value work." The implementation begins.

But in my work securing these systems and consulting on their deployment, I've observed something that rarely makes it into the success metrics: the human cost of these transitions.

In the introduction, I shared with you the incident early in my career when I misinterpreted a colleague's legitimate medical research as inappropriate web browsing. That experience taught me to look for the human context behind the data. But as systems become more automated, who's looking for that context? Who's ensuring that efficiency doesn't erase what matters to people? Will we even be in a position to look for this context in the future?

This isn't about technology being evil. It's about what happens when we optimize for everything except human meaning and purpose. I've seen it enough times now — those slight changes in workplace culture, the way conversations shift, and the gradual loss of trade craft and institutional knowledge.

In cybersecurity, specifically, I've watched as it has become more and more automated. The algorithms get smarter, faster, and more comprehensive. As it has taken the human element out of the process, I've watched hackers exploit the weaknesses and slip through. Cybersecurity was never just about stopping threats; it was about understanding human behavior and recognizing patterns that might indicate someone under duress or making desperate choices. It's appropriate risk management, and much of that centers around the human risk and the human attack surface. There is wisdom in the human element of cybersecurity that doesn't easily translate into code.

The pattern isn't limited to any single industry. It's the same consulting companies using their version of the same template applied across sectors: identify inefficiencies, implement technological solutions, measure improvements, and declare success. However, the human elements — such as the sense of purpose, the feeling of being needed, and the accumulation of expertise — don't fit neatly into performance dashboards.

What concerns me most is the gap between the metrics that show "improvement" and the reality of human experience within these transformed systems. The numbers might look good, but people are telling a different story in break rooms, in support groups, and in the quiet moments when they wonder if their years of experience still matter.

This observation has shaped my approach to every technological implementation, not with blind resistance, but with persistent questions: What are we optimizing for? Who decides what constitutes improvement? What essential human elements might we be overlooking?

The technology itself isn't the problem. The problem is how we implement it without considering these deeper human impacts. And that's a choice we make, or fail to make, every single day.

As we stand on the brink of even more transformative technological changes, these patterns serve as a warning. Without intentional design that preserves human dignity and purpose, we risk creating a world where an increasing number of people wake up wondering if anyone truly needs them anymore.

And that question, "Am I still needed?" might be one of the most dangerous questions our society faces.

Industrial Ghosts Still Haunt Us

The factories are gone.

Their concrete foundations have crumbled. Their iron beams have rusted. Their empty parking lots have sprouted weeds through cracked asphalt. But their ghosts linger among us.

They whisper warnings we still refuse to hear.

Every time we choose efficiency over humanity, metrics over meaning, and scale over community, we summon them. Every time we speak of "human resources" or "human capital" or "optimization of headcount," we echo the language that once reduced craftsmen to cogs, and artisans to appendages. Every time we design systems that treat people as problems to solve rather than souls to nurture, we lay another brick in the road haunted by these ghosts.

I remember lessons from my past when I listened to tech executives extol the coming wave of AI. I hear echoes of past promises:

"This will create new opportunities." "This will free people for more creative work." "This will drive progress."

Perhaps. But it may also unleash forces that we cannot control, because the machines are no longer confined to factory floors and assembly lines. They're embedded in our daily rhythms, nestled in our pockets, humming in our homes. They're learning our habits, anticipating our desires, and shaping our attention in ways we barely perceive. They've become intimate with us in ways industrial machinery never could.

The danger today isn't just economic displacement. It's psychological dissolution: a world where human worth is increasingly measured by how well we integrate with systems designed for speed, scale, and profit instead of meaning, connection, and purpose.

We risk becoming strangers to our own worth.

But here's what haunts me most deeply: Meaning doesn't disappear quietly when we strip it away. It doesn't dissolve into resignation or acceptance. It transforms.

Into resentment. Into despair. Into rebellion.

We witnessed this during previous industrial transitions: the rise of extremism in hollowed-out communities, the surge in deaths of despair, the fracturing of social cohesion. We're seeing early tremors again today: polarization that defies rational explanation, tribalism that overwhelms shared truths, and a desperate grasping for identity and belonging in all the wrong places.

These aren't glitches in our social system. They're features of a world that has forgotten the human at its center.

So, where do we go from here?

We remember.

We remember that efficiency is a means, not an end in itself; that optimization is a tool, not a value in itself; and that progress without purpose is merely motion, not advancement.

We remember that people are not data points to be managed, resources to be allocated, or problems to be solved. We are messy, contradictory, glorious beings: wired for connection, meaning, and contribution beyond what can be measured in productivity dashboards.

If we build systems that forget this essential truth, no amount of technological marvel will save us from what follows.

I sometimes imagine the ghosts of those old factory workers watching us now: leaning against crumbling pillars, arms folded across faded work shirts, eyebrows raised in recognition as we race toward a cliff they've already fallen from.

Will we learn from their fall? Or will we, dazzled by the promise of progress, accelerate toward the same abyss?

I want to believe we can choose differently and build a world where technology amplifies humanity rather than replacing it. I want to believe we can design progress with purpose, care, and community at its core.

But that possibility depends on our willingness to do one difficult thing: remember.

Remember the workers whose craftsmanship was deemed dispensable. Remember the communities hollowed out by efficiency. Remember that every technological revolution carries human consequences that spreadsheets can't capture.

The industrial ghosts still walk among us. They are not asking us to turn back the clock or halt innovation. They are simply asking us to remember what they learned too late:

When we forget the humans, we destroy the very purpose of progress itself.

Chapter 3
The New Invisible Worker

The Promise and the Betrayal

For a long time, the office was seen as an escape.

It was where you went to leave behind the grease-stained overalls and the deafening clang of the factory floor. It was the promise of a world where the mind mattered more than the muscle, where success wasn't measured by how many pounds you lifted or how many widgets you assembled, but by the clarity of your thinking, the brilliance of your ideas, and the depth of your judgment. It was the return of craftsmanship, just with words instead of wood, solutions instead of metal, strategies instead of seams.

At least, that's what we believed.

In the summers of my late teenage years, I worked for an underground contracting company. Which means I dug ditches all summer. I was the grunt. I remember my early days stepping into an office environment, feeling like I had arrived. Coming from a background where I was more used to blue-collar grit than white-collar polish, it felt almost like cheating to get paid to think. I was intoxicated by the feeling that ideas mattered, imagination had value, and people were willing to pay for what was in my head instead of what I could hammer together with my hands.

But here's the hard truth I wish I had seen back then: Technology doesn't stop at the factory gates. It slips under the office door, winds its way around cubicles, and becomes embedded in every process, interaction, and expectation. What began as tools to support human ingenuity slowly morphed into systems that demand we keep up with them. And somewhere in that shift, something precious began to erode.

We told ourselves that automation was for the factory worker, punch-card employee, and person on the assembly line whose hands could be replaced by a robot's. But now I see it clearly — knowledge work was never immune from automation. We just didn't recognize the threat until it was too late.

When your work becomes a commodity and that work is reduced to a set of tasks, you become disposable.

The Rise of the Invisible Class

It didn't happen overnight.

The fall of the knowledge worker, or maybe more accurately, the hollowing out, has been a slow, almost imperceptible process. Like a house that looks sturdy on the outside but is rotting from within, the walls have been quietly crumbling for years.

We just didn't want to see it.

For decades, we white-collar workers believed we were immune. We watched automation replace factory workers, cashiers, warehouse staff, bank tellers, and we told ourselves that our realm was safe. After all, you can't automate creativity. You can't outsource judgment. You can't write an algorithm for wisdom.

Or so we thought.

But the digital revolution is nothing if not thorough. It doesn't stop at the places we think are off-limits. It burrows into every space where efficiency can be extracted.

I've watched AI creep into the roles of analysts, writers, graphic designers, and even therapists. I've seen customer service systems that no longer need human agents except for the most complex calls. I've seen financial analysts displaced by predictive software. And each time, there's this moment, this collective flinch, when people realize: *Oh. This isn't just about blue-collar work. This is about all of us.*

Automation, once a wave hitting the factory floor, has surged into the office tower. It's no longer about replacing muscle. It's about replacing minds.

The Impact

Let's break it down.

- Automation of white-collar tasks: What used to be complex, integrated work is now sliced into micro-tasks, discrete chunks that can be handled by machines or outsourced platforms. Writing isn't about thought leadership; it's about generating SEO content. Design isn't about creativity; it's about churning out deliverables to feed the content beast. Analysis isn't about nuance; it's about dashboards spitting out insights in seconds.
- Remote and gig work: The rise of remote work and the gig economy has fragmented workers in ways we're only beginning to understand. On paper, flexibility sounds like liberation. But in practice, it can mean isolation, uncertainty, and relentless competition. You're no longer a valued team member; you're a line item on an outsourced contract, a replaceable node in a global talent market.
- Digital surveillance: Every click, keystroke, and calendar entry are now measured, analyzed, and optimized. Productivity

software promises to help us do more, faster. But what it often does is transform workers into data points, stripping away autonomy, creativity, and the rhythms that make us fully human.

How Office Work Is Being Eroded

You don't wake up one day and find your job is now meaningless. It happens in fragments — slow, invisible shifts that eat away at purpose.

We often envision collapse as a dramatic event: a company folding overnight, a mass layoff announcement, or an industry-wide upheaval. Yet, the real erosion of office work is quieter, subtler. It happens under the surface, like a foundation slowly cracking beneath the facade.

Let's peel back the layers.

Task Disintegration: Sliced and Automated

Once upon a time, work had a shape.

It was messy, yes, full of judgment calls, unpredictable turns, and creative detours. But it was whole. You could point to something at the end of the day and say, *I made that. I solved that. I shaped that.*

Now? Work is becoming progressively fragmented.

AI systems, automation platforms, and digital workflows break down work into micro-tasks. Instead of designing an experience, you're tweaking button placements. Instead of writing a strategy, you're generating endless A/B tests. Instead of mentoring a team, you're approving help-desk tickets.

I remember when I first noticed this in my own work. What used to be a flowing, intuitive process became checklist driven. Tools promised to "help," but they started to dictate my thinking. I wasn't shaping the work anymore; the work was shaping me.

The tragedy isn't just lost craftsmanship. It's lost coherence. You no longer see the arc of what you're building, only the fragments you're managing.

Identity Loss: When Jobs Stop Meaning Anything

For centuries, work has been tightly bound to identity.

We ask children, *"What do you want to be when you grow up?"* not *"What do you want to do?"* We define ourselves by job titles, industries, and missions.

But when work becomes fragmented, so does identity.

A developer no longer feels like a builder; they are a ticket closer. A designer no longer feels like a creator; they are a generator of assets. A manager no longer feels like a mentor; they are a metric driver.

I've sat with people — smart, talented, deeply committed people — who've told me they no longer know who they are at work. The job still pays. The calendar is still full. The LinkedIn profile still sparkles. But when they look in the mirror, they see the person slipping away.

Social Erosion: The Crumbling of Connection

I'll say something that's almost taboo in the age of remote work: We are starving for human connection.

Remote and hybrid work have opened beautiful possibilities, but have also thinned out the social fabric of work.

Mentorship happens less often. Serendipitous conversations happen less often. The small, quiet rituals that make teams feel like teams — coffee runs, hallway jokes, and after-hours debriefs — occur less frequently.

The remote work environment has created new patterns of professional isolation. People can work for years alongside colleagues they interact with only through screens and messages. They can maintain all the professional connections, attend all the meetings, respond promptly to messages, and complete projects on time, all while missing the informal human moments that once built workplace relationships. The efficiency of digital communication can coexist with profound loneliness, creating a disconnect between being connected and feeling seen.

There are people I've worked with in the past whom I would not be able to recognize on the street because I've never actually seen their faces.

This is the erosion we don't talk about enough — not just the erosion of tasks or roles, but the erosion of belonging.

Warning Signs We're Missing

If you walk into most offices today, virtual or physical, you probably won't see chaos. You'll see dashboards glowing with metrics. You'll hear leaders celebrating "record productivity." You'll see headlines marveling at how AI is unlocking untapped potential.

On paper, things look great.

But beneath the surface, something is breaking.

We are so busy measuring output that we are missing the collapse of purpose.

Let me walk you through the warning signs — the ones that don't show up on a quarterly report but that tell us everything we need to know.

Burnout Despite Automation

Automation was supposed to free us.

With technology handling the drudgery, we were told, we'd have time for deep work, creative thinking, and big-picture strategy. We'd finally get to be the best versions of ourselves at work.

Instead? Burnout is everywhere.

Why?

Because automation doesn't reduce demands. It raises expectations.

You automate one task, and three more appear. You speed up one workflow, and the deadline becomes even tighter. You increase capacity, and the bar moves higher.

The paradox of modern technology is increasingly clear: We've never been more efficient, yet burnout rates continue to climb. Automation and optimization tools promise to save time, but the reality is often quite different. The same technology that should free us up instead creates new expectations for constant availability and accelerated output. This isn't a coincidence - it's the fundamental tension of a workplace designed around machines rather than humans.

Efficiency without meaning doesn't create freedom. It creates emptiness.

Ironically, it doesn't seem to create profitability either.

Quiet Quitting and Disengagement

There's a reason "quiet quitting" has become a prominent topic in the cultural conversation following the pandemic.

It's not laziness. It's not entitlement. It's not a generational failing.

It's a signal.

When work feels disconnected from purpose, and people are treated as throughput engines rather than human beings, they withdraw. Not always dramatically. Sometimes just ... quietly.

They stop going above and beyond. They stop raising their hand in meetings. They stop dreaming about what's possible.

They're still there, but something inside has gone dark.

I've seen this up close, and it's haunting. You can have a room full of brilliant, talented people, and if they no longer believe in the "why," you have a hollowed-out team that's just silent.

Skill Decay: From Creators to Operators

One of the least talked-about consequences of automation is skill decay.

As we hand over more tasks to machines, we risk dulling the very abilities that made us valuable in the first place.

I think about the young analysts who will no longer learn how to interpret data because the dashboard spits out the insights. I think about the designers who will stop practicing the fundamentals because templates do the heavy lifting. I think of all those in whatever field they are in who will lean so heavily on these new tools that they start to lose the muscle of original thought.

We are not just automating work. We are automating away the growth of human expertise.

I fear the more passive we become, the more fragile we become.

The Future of White-Collar Work

So here we are once again. We are in a moment that feels both familiar and new.

On the one hand, this isn't new for the average white-collar knowledge worker. We've seen waves of transformation efforts sweep through industries, displacing workers and reshaping economies. We've seen progress carry both promise and danger.

But this time is different.

Now it's not just our hands that are being replaced, it's our minds. It's not just our physical labor being automated; it's our thinking, our creating, our judging, and our imagining. And that means the question we face is not just "What do we do?" It's also "Who do we become?"

We Won't Survive by Competing with Machines

Here's the temptation: We try to outrun the machine. We strive to match its speed, efficiency, and precision. We try to become faster, sharper, leaner, until we are no longer recognizable to ourselves.

But that's a short-sighted losing game, because the machine will always be faster. The algorithm will always process more data. The AI will always generate more variations, more iterations, and more output.

Our survival does not lie in outcompeting the machine. It lies in becoming more human.

The Return to Human Strengths

The future belongs to those who can amplify what machines cannot replicate:

- Creativity — not just generating options, but imagining what's possible.
- Empathy — not just reading emotions, but responding with care.
- Judgment — not just choosing between A and B, but knowing when to choose neither.
- Wisdom — the slow, earned ability to understand complexity, nuance, and meaning.

Machines can do amazing things. But they cannot love. They cannot grieve. They cannot hope. They cannot hold a friend's hand and say, "I see you."

That's where our value lies. That's where our resilience lives.

A Personal Challenge to the Reader

So, I want to ask you, not as a distant author, but as a fellow traveler:

Where in your life have you already started slipping into the automation trap? Where have you traded depth for speed, meaning for metrics, presence for productivity?

And more importantly, where can you reclaim your humanity?

Maybe it's in how you lead your team. Maybe it's in how you mentor the intern. Maybe it's in how you approach your craft — not as a series of tasks, but as an expression of something only you can bring.

Whatever it is, don't wait. Start now because the machines are not slowing down. But neither is the human spirit.

Part Two
The Human Cost

Chapter 4
The Risk of System Slaves

Freedom in Disguise

Freedom today wears a disguise so elegant that we sometimes forget it's a mask at all.

We're told we're free because we can choose among endless options – a thousand brands filling digital shelves, countless streaming options awaiting our selection, an infinite scroll of content tailored to our preferences. We're assured of our autonomy because we can work from anywhere. Our laptops have become portable offices that transform coffee shops and kitchen tables into corporate outposts. On the surface, it's dazzling: a tapestry of choice, flexibility, and convenience that would leave our ancestors speechless with wonder.

Yet beneath this glittering surface lies a harder truth I've been reluctantly facing: The system defines the choices, establishes the terms, and subtly punishes those who step outside its invisible boundaries.

I remember when this realization first settled into my consciousness like a stone dropping through still water. I was at a cabin in Gatlinburg, Tennessee, during what should have been a complete disconnection from work, my first real family vacation in three years. I sat on the cabin's deck as the morning light filtered through the Smoky Mountain trees, and yet my brain kept telling my hands to reach toward my phone to check my messages. It was a subconscious reflex that I couldn't stop or explain.

Check emails. Review analytics. Respond to that thread.

I was on vacation, and no one was expecting me to respond to anything. No one had asked me to stay connected. There was no external pressure, and yet I felt compelled, as though invisible strings were pulling at my fingertips. That's when I understood: I hadn't escaped the system at all. I had transported it with me, hundreds of miles from the office, tucked neatly into my pocket.

In the industrial age, the chains of control were visible and tangible. You could see the factory walls, hear the foreman's commands, and feel the physical exhaustion in your muscles and bones. Timecards and factory whistles marked the boundaries between work and freedom.

Today, these boundaries have dissolved into something far more insidious: psychological tethers, algorithmic nudges, and most dangerously, self-imposed limitations. We've internalized the system so completely that external enforcement has become redundant.

I see this silent transformation everywhere I look:

- Employees no longer compete for pride in their craft but begrudgingly obsess over performance metrics that are detached from meaning.
- Professionals network not to build meaningful relationships but to enhance their profiles for recommendation engines and career advancement systems.
- Even our leisure moments are shaped by systems that harvest our preferences, monetize our attention, and subtly redirect our desires.

We aren't forced into obedience through threat or coercion. We're gently nudged, cleverly gamified, continuously scored and ranked until compliance feels like the only logical choice. The path of least resistance becomes a rut so deep that we mistake it for the only path available.

The true genius of the modern system, what makes it so devastatingly effective, is that it convinces us that submission is, in fact, freedom.

Want to take a break? The notification reminds you that you haven't closed your activity rings. Want to express yourself authentically? The platform quietly buries content that doesn't fit engagement parameters. Want to stand apart? The algorithm offers a selection of pre-approved "uniqueness" that never threatens the underlying structure.

I've sat with this uncomfortable truth for years now, turning it over like a pattern I can't unsee once I've noticed it: The most dangerous prison may be the one with walls so beautiful that we hang pictures of them in our minds and call it freedom.

And that, I believe, is the world we inhabit now.

Digital Chains

It didn't happen in a single transformative moment.

There was no grand announcement, no day when we collectively agreed to surrender our autonomy to code and metrics. No authority figure stood at a podium and declared, "From today forward, you will be governed by algorithms you cannot see, systems you cannot question, and values you did not choose."

Instead, it arrived in whispers and conveniences, a gradual accumulation of small surrenders that seemed insignificant in isolation: A notification that interrupted a moment of reflection. A productivity app that promised to optimize our time. A feedback system that reduced complex human interactions to star ratings. A recommendation algorithm that narrowed our horizons while promising to expand them.

I remember watching this shift happen within my own professional life. Early in my career, leadership meant direct human connection –

conversations in hallways, handwritten notes, the subtle reading of expressions in a meeting room. Twenty years later, I found myself hearing discussions of measuring team members in terms of their dashboard metrics, making decisions based on data visualizations rather than conversations. The change was so gradual I barely noticed it happening.

What would once have been recognized as control mechanisms were rebranded with terminologies that felt empowering rather than restrictive: Command-and-control became "seamless workflows." Surveillance transformed into "performance insights." Conformity pressure was renamed "best practices."

The reality of algorithmic management is well-documented across numerous studies and reports. Delivery drivers, ride-share operators, and other platform workers operate under systems where algorithms, not humans, determine their routes, schedules, compensation, and performance evaluations. The opaque nature of these systems leaves workers with little recourse or ability to question decisions that directly impact their livelihoods.

This shift represents something fundamental: the transfer of management from human supervisors who could be questioned, reasoned with, or appealed to, toward algorithmic systems that operate according to invisible rules. Workers report feeling they must "play the game" according to parameters they don't fully understand and cannot negotiate with.

What's most concerning isn't just the control itself, but the resignation that comes with it, the widespread acceptance that this is simply "how things work now." The choice becomes binary: adapt to the algorithmic overlords or lose your income. There is rarely space for questioning the system or reimagining how technology can serve human needs rather than the reverse.

This pattern extends beyond delivery apps and ride-sharing services into countless workplaces, where algorithms increasingly determine workflow, productivity targets, and even hiring and firing decisions. The relationship between humans and systems has undergone a fundamental shift, with technology evolving from a tool to a taskmaster.

The transition from physical to digital chains has happened so seamlessly that we barely noticed as the architecture of control migrated from external authorities to internal compulsions, and from visible structures to invisible nudges that shape our behavior from within.

Today, the human spirit faces a test unlike any in our history: Will we recognize the gradual erosion of our autonomy, character, and moral agency? Or will we, seduced by the undeniable conveniences and efficiencies of our digital systems, continue to tighten the chains ourselves, mistaking them for bracelets?

When Systems Manage Humans

There was a time, not so distant in our collective memory, when the relationship between humans and machines was clearly defined. We built tools to serve our needs, established the parameters of their operation, and determined what constituted value and meaning.

But somewhere along this technological journey, a subtle inversion occurred. The relationship flipped, and we barely noticed as it happened.

More and more often today, it's the machines and the systems behind them that govern us. This isn't a dystopian fantasy or a futuristic concern. It's the quiet reality reshaping our daily lives, our work, and our sense of agency in the world.

Algorithmic Management

The modern workplace has undergone a transformation that few of us fully comprehend, even as we live within its structures. Particularly in the expanding realm of the gig economy, workers aren't merely supervised by other humans: They're managed, measured, and evaluated by algorithms.

I witnessed this firsthand when I was a consultant and had the opportunity to tour a fulfillment center where employees' movements were tracked with a precision that felt almost surgical. How many items were picked per hour? How many steps were taken between tasks? How many seconds were spent in transition? How often did they deviate from optimal pathways?

The human supervisors I've spoken with in these environments often don't make decisions. They simply deliver the assessments the system hands them. Some of them were unaware of that fact until I pointed it out. The algorithm doesn't account for individuals with unique circumstances, physical limitations, or creative problem-solving abilities. It registers only deviations from the optimized pattern.

Major e-commerce fulfillment centers utilize sophisticated tracking systems that monitor workers' every movement through handheld scanners and wearable devices. The "Time Off Task" (TOT) system automatically tracks any period when workers aren't actively scanning items. According to documents revealed in labor board proceedings and investigative reporting, the system flags workers who accumulate too much TOT, which can include bathroom breaks, brief conversations with colleagues, or moments to rest. Technology news outlets reported in 2019 that automated systems in some warehouses can generate warnings and even termination paperwork without human manager involvement, based purely on productivity metrics. Workers have reported receiving disciplinary actions for time spent walking to distant bathrooms or addressing equipment malfunctions — circumstances the algorithm doesn't distinguish from intentional work avoidance.

Ride-sharing platforms utilize sophisticated behavioral psychology through their driver apps, as documented in research published in academic journals, including the *University of Chicago Law Review*.[1] The apps use various "nudges," including surge pricing heat maps that create visual stimuli designed to influence driver behavior. Studies have shown the apps employ gamification techniques such as earning streaks, achievement badges, and forward dispatch (automatically queuing the next ride before the current one ends).[2]

Research by scholars like Alex Rosenblat[3] has documented how these platforms utilize psychological techniques borrowed from video game design to keep drivers on the road longer. These techniques include progress bars that fill as drivers near arbitrary earning goals and notifications that appear just as drivers attempt to log off, suggesting they're close to hitting bonus targets.

Modern call centers are steadily deploying emotion AI and voice analytics software to monitor not just what agents say, but also how they say it. Voice analytics companies provide real-time voice analysis to give agents on-screen coaching about their emotional tone and delivery.

According to reports by major financial publications and tech industry analyses[4], these systems measure factors such as speaking pace, tone variation, and perceived enthusiasm levels. The software can generate real-time prompts telling agents to "speak more slowly," "add energy to your voice," or "show more empathy." Performance evaluations are incorporating these emotional metrics alongside traditional metrics, such as call duration and resolution rates. Research from Cornell University[5] has documented how these systems can detect micro-expressions in voice patterns, creating unprecedented levels of emotional surveillance in the workplace.

What's most unsettling isn't the efficiency or control, but the way these systems reshape how people understand themselves. When you are constantly reduced to a metric, a scan rate, a delivery quota, or a customer rating, you begin to internalize that message. You stop seeing yourself as a creative, adaptive human and start seeing yourself as a performance variable.

And that's the deeper cost of algorithmic management: It doesn't just regulate behavior; it reshapes identity. It erodes the sense of agency, meaning, and worth that make work and life human.

Social Media Control

This management by algorithm extends far beyond the workplace.

Social media platforms don't merely reflect human interactions; they actively shape them through invisible incentives and penalties.

What content appears in your feed? Which posts receive amplification? Which emotions generate engagement? Which perspectives are deemed relevant?

I've watched friends and family members unconsciously adapt their expression to fit the reward systems of various platforms: choosing topics, tones, and even emotional registers based on what generates validation through likes and shares. I've caught myself doing the same. You probably have too — that momentary hesitation before posting something authentic but potentially "unpopular" with the algorithm.

Popular short-form video platforms utilize recommendation algorithms that, according to company documents and researchers, are among the most sophisticated content curation systems ever developed. Major news outlet investigations revealed that these algorithms can determine user interests with remarkable precision within hours of initial use. Academic studies have documented how content creators report changing their content style, appearance, and even personality to match what generates algorithmic promotion.

Research from the Oxford Internet Institute[6] found that creators often experience what researchers' term "algorithmic anxiety" — stress and uncertainty about how to maintain visibility as platform algorithms evolve. Multiple studies have shown that these powerful recommendation systems can create "filter bubbles" more quickly and comprehensively than other social media platforms, fundamentally shaping not only what users see but also what creators feel compelled to produce.

Major professional networking platforms make use of the "Social Selling Index" (SSI) scoring, which quantifies professional networking, fundamentally altering how professionals approach their career development. The platform's algorithm allegedly assigns users a score ranging from 0 to 100 based on four components. According to platform data and independent analyses, recruiters and hiring managers reference these scores when evaluating candidates.

Research published in business journals[7] has documented how the gamification of professional networking has led to what scholars call "performative professionalism" — behavior oriented toward algorithmic metrics rather than authentic professional relationships. The pressure to maintain high scores has created a new form of digital labor, where professionals feel compelled to regularly post content, engage with others' posts, and expand their networks in ways that may feel inauthentic but are necessary for career advancement.

Research has consistently shown that photo-sharing platform engagement algorithms favor certain types of visual content, inadvertently establishing and reinforcing specific beauty standards. A study published in the journal *Body Image*[8] found that the most-promoted content on major visual platforms typically features specific body types, skin tones, and facial features, creating a narrow definition of beauty.

The widespread use of beauty filters has created what psychologists term "filter dysmorphia" — a phenomenon where people seek cosmetic procedures to look more like their filtered selves. According to research from Boston Medical Center[9], plastic surgeons report a significant increase in patients bringing filtered selfies as reference images for desired procedures.

Studies have also documented[10] how platform algorithms tend to promote content that receives high engagement quickly, creating a feedback loop where certain aesthetic standards become increasingly dominant. In contrast, diverse representations of beauty become marginalized in users' feeds.

Our attention isn't just captured; it's cultivated. Our emotional responses aren't just observed; they're harvested. Our social connections aren't just facilitated; they're engineered.

We are no longer simply using social media — social media is using us, reshaping our communication patterns, our attention spans, and even our perception of reality itself.

Predictive Profiling

Perhaps most concerning is how prediction technologies are redefining human potential before it has the chance to express itself.

We are now frequently defined not by who we are or even by what we've done, but by what algorithms predict we might do: your credit-worthiness score. Your health risk profile. Your "fit" for a particular job or opportunity. Your projected lifetime value as a customer.

Medical insurance companies increasingly use predictive analytics that incorporate data far beyond traditional health records. According to investigative journalism and academic research, insurers purchase and analyze consumer data, including retail purchase history, credit card transactions, social media activity, and geographic data. These algorithms can factor in seemingly unrelated variables, including shopping patterns at specific stores, gym membership data, and even social media posts, to predict health risks and set premiums.

Studies have documented cases[11] where ZIP codes serve as proxies for race and socioeconomic status, potentially leading to discriminatory pricing. Research from health policy journals reveals that these predictive models often create feedback loops where those deemed "high risk" face higher costs, potentially pricing them out of adequate coverage and creating the very health disparities the algorithms claim to predict.

These predictions don't just describe possibilities. They actively create them by opening or closing doors before we even reach them.

Educational institutions are more and more often employing predictive analytics to forecast student success and determine academic pathways. Research published in educational technology journals documents how these systems analyze variables, including standardized test scores, attendance patterns, demographic data, and even library usage to predict academic outcomes.

A comprehensive study by the National Education Policy Center[12] found that many of these algorithms inadvertently perpetuate existing inequalities by identifying patterns that correlate with socioeconomic status rather than actual academic potential. These systems often create what researchers call "algorithmic tracking," where students are sorted into different educational paths based on predictions that become self-fulfilling prophecies.

Documentation from several school districts[13] shows how these predictions influence resource allocation, teacher assignments, and enrichment opportunities, potentially limiting students' futures based on algorithmic assessments made early in their academic careers.

A landmark bias study by a major tech company[14], reported by news agencies, found that their AI recruiting tool systematically downgraded resumes containing words associated with women's colleges or women's activities. The Proceedings of the National Academy of Sciences demonstrated that AI hiring systems often perpetuate historical biases present in training data. Studies have shown these systems[15] can discriminate based on names, ZIP codes, educational institutions, and even word choices that correlate with protected characteristics.

Research from MIT[16] found that video-interview AI systems show bias against non-native English speakers and individuals with specific facial features or expressions. Business journals report[17] that many companies struggle to justify their AI hiring decisions due to the "black box" nature of machine learning algorithms, making it impossible to identify or correct potential discriminatory patterns.

The algorithms themselves weren't malicious; they had simply optimized for patterns in the existing workforce. However, in doing so, they systematically closed doors for qualified candidates who took non-traditional paths, all without human oversight, until they specifically looked for this pattern.

This is the core insight of system slavery today: Humans are no longer primarily managing machines. Machines are managing humans. And this inversion comes with profound costs: economic, psychological, moral, and existential.

When algorithms decide what matters, we gradually forget that we ever had the authority to choose differently.

The Psychology of System Slavery

Having explored the external architecture of control, we must now turn inward to examine something equally important: the internal transformation that occurs when systems reshape not just our environments but also our very sense of self.

This is the psychological dimension of system slavery, perhaps the most profound and least discussed aspect of our technological age.

Learned Helplessness

In the late 1960s, psychologist Martin Seligman conducted landmark research[18] on a phenomenon known as "learned helplessness." When subjects repeatedly experienced conditions they couldn't control, they eventually stopped trying to change their circumstances, even when opportunities for escape became available.

Not all test subjects (animals) in the experiment exhibited the same pattern of passive acceptance. I believe this to be true among humans as well. This is something we will discuss in the next chapter. In the meantime, let's explore the idea of learned helplessness a little further.

I've heard stories of this phenomenon occurring with disturbing frequency in certain environments. Consider the customer service representative who follows a nonsensical script even when it clearly isn't serving the customer. Think about the teacher who abandons proven educational approaches to satisfy standardized metrics. Or reflect on the healthcare provider who rushes through patient interactions to meet throughput quotas. In each case, they seem to reveal the same pattern: initial resistance, followed by persistent frustration, and finally, a resignation that masks deeper despair.

Over-Adaptation

Learned helplessness is not the only way we adapt under pressure. Sometimes, instead of shutting down, our nervous system swings in the opposite direction, becoming hyper-attuned to what is overwhelming us. The body, which has become inadvertently trained to expect constant interruption, begins anticipating signals that aren't there. It's a different manifestation of the same underlying problem: systems that reshape not just our behavior, but our very biology.

Phantom Vibration Syndrome has been documented in numerous medical studies[19] as a form of tactile hallucination where individuals perceive their phone vibrating when it hasn't. Research published in the *British Medical Journal* and other peer-reviewed publications found that up to 89% of college students and 68% of medical professionals experience this phenomenon.[20] Neurological studies have shown that constant device use can create new neural pathways that anticipate notifications, leading to false sensory perceptions.

Brain imaging research has documented[21] actual changes in somatosensory cortex activity among individuals who frequently use technology. The condition represents a physical manifestation of psychological dependence on digital devices, with studies showing a correlation between phantom vibrations and anxiety levels, device attachment, and occupational stress. Medical researchers have termed this a form of "learned bodily habit" where the nervous system adapts to expect constant digital interruption.

Stanford University researchers identified video-call fatigue as a genuine psychological phenomenon with specific neurological causes. Their studies, published in *Technology, Mind, and Behavior*[22], identified four primary causes: excessive close-up eye contact, seeing yourself constantly, reduced mobility, and higher cognitive load from interpreting non-verbal cues. Research has shown that video calls require more mental processing than in-person interactions, as the brain works harder to interpret facial expressions and body language through a screen.

Studies have documented[23] how the lack of natural conversational rhythms and the inability to use peripheral vision for social cues create cognitive overload. The phenomenon became particularly acute during the pandemic, when knowledge workers reported spending eight to 10 hours daily in video meetings. Neuropsychological research has shown that this constant video interaction can lead to decreased attention span, decision fatigue, and what researchers term "nonverbal overload" — exhaustion from over-interpreting limited visual cues.

This learned helplessness doesn't happen overnight. It accumulates through countless moments of having your judgment overridden, your expertise ignored, and your humanity reduced to data points. Eventually, the path of least resistance becomes not just compliance, but the surrender of the very idea that things could be different.

I've seen this transformation in many professions — people who once approached their work with passion and creativity gradually reduced to mechanically executing protocols they know are harming the very mission they once believed in.

What's most heartbreaking isn't watching their spark fade, it's recognizing how deeply they've come to believe that resistance is futile. One person put it to me this way: "What's the point of fighting it? The numbers are all that matter to anyone now."

The system hadn't just shaped their behavior; it had trained them to stop imagining alternatives, and it reshaped their sense of agency. They no longer saw themselves as active participants, but as passive operators of a system they no longer believed in and could no longer imagine changing.

Depersonalization

As systems reduce complex human realities to quantifiable metrics, something profound happens to our self-perception: We begin to see ourselves, and others, through the system's eyes.

I've seen this happen repeatedly in the world of security awareness training.

What began as a genuine effort to help people navigate an increasingly dangerous digital world has, in many places, been reduced to a compliance exercise. Training programs are implemented not because they benefit people, but because they check a box on an audit report.

I hear it from teams all the time:

- "Did everyone complete the training?"
- "What's our failure rate on the phishing test?"
- "Are we below the industry benchmark?"

And here's the heartbreak. Is anyone asking:

- "Did our people actually understand why this matters?"
- "Do they feel more confident, more empowered?"
- "Are they coming to us when they spot something suspicious, not out of fear, but because they trust we've got their backs?"

Instead, we've built dashboards where the only meaningful categories are "Pass" and "Fail." We've built cultures where employees rush through videos at 2x speed just to get through them. The real lesson they learn is not how to be safer, but how to avoid getting in trouble.

We turn people into risk scores. We turn learning moments into punishment cycles. And slowly, we erode the most important part of a security culture: the human connection.

I've asked leaders, "What's the point of your security awareness program?" Too often, the honest answers are in this order:

1. "To avoid fines."
2. "To satisfy regulators."
3. "To get the auditors off our backs."

But security is not a checkbox. It's a conversation. It's a relationship between an organization and its people, one built on trust, transparency, and the shared goal of protecting what matters.

When you reduce security awareness to a box on a form, you're not just failing your security program. You're failing your people.

And here's the real danger: When people feel like cogs in a compliance machine, they stop caring. They stop asking questions. They stop believing that they matter in the fight to keep the organization safe.

And that, more than any phishing email, is what puts an organization at risk.

This is depersonalization in action, the gradual erosion of the sense that you are a whole, complex, and irreducible human being. It represents a profound identity shift from "I am a person who does work" to "I am a work function that just happens to reside in a person."

The implications of this shift are significant and cannot be overstated. When you no longer recognize yourself as fundamentally human, with all the messy complexity that entails, it becomes easier to accept treatment that diminishes that humanity. The system's logic becomes your logic. Its values become your values. Its definitions of success and failure become the lens through which you evaluate your worth.

Moral Fatigue

Perhaps the most disturbing psychological consequence is what some refer to as "moral fatigue," the exhaustion that comes from constantly navigating system demands that conflict with human needs and ethical intuitions.

I see this in nearly every sector: The customer service agent is measured solely on call handling time rather than problem resolution or customer satisfaction.; The journalist is incentivized to produce viral content rather than substantive reporting. The financial advisor is pressured to recommend products that generate fees rather than benefit the client.

These aren't bad people making selfish choices. They're good people caught in systems that make ethical action more costly.

At first, they resist. They find workarounds. They try to protect what matters while still surviving within the system.

But over time, as the pressure remains relentless, something gives way. The constant friction between what feels right and what the system demands creates a unique kind of exhaustion that extends beyond physical or mental fatigue. It erodes the will to fight, the capacity to care, and the energy to imagine alternatives.

Eventually, the questions change. People stop asking "What's right?" and start asking "What's required?" They stop pushing back and start going through the motions, not because they've become different people, but because the system has gradually worn down their capacity for moral resistance.

Content creators across platforms face what researchers have termed "algorithmic gaslighting" — the psychological impact of constantly changing platform algorithms that alter content visibility without warning or explanation. Studies by digital media researchers have documented how changes to platform algorithms can instantly devastate creator revenues and audience reach. Creators on major video platforms have reported 90% drops in viewership following algorithm updates, as documented in research from the University of Amsterdam[24].

This instability creates what psychologists studying digital labor refer to as "platform precarity" — a state of constant uncertainty about the rules that govern one's livelihood. Research published in *New Media & Society*[25] found that creators experience symptoms similar to those found in emotionally abusive relationships: self-doubt, anxiety, and a constant questioning of their perception of reality. The opacity of these algorithms means creators can never be certain whether declining performance reflects their content quality or invisible algorithmic changes.

This isn't the dramatic moral collapse portrayed in fiction. It's quieter, slower, and more mundane – and in that subtlety lies its power. Like erosion shaping a landscape over time, moral fatigue gradually reshapes our sense of what is possible, what is worth fighting for, and ultimately, who we are.

How We Become Complicit

Here we reach what may be the most difficult truth in this exploration: The system doesn't merely act upon us as passive recipients. It works through us. We become, often unconsciously, its agents and amplifiers.

We are not just victims of system slavery. We are, in ways both subtle and profound, its collaborators.

This isn't about assigning blame. It's about recognizing a dynamic that will continue unless we consciously interrupt it. And the first step is understanding how this complicity takes root.

Prioritizing Metrics Over Meaning

It begins innocuously, with the simple human desire to succeed within the parameters we're given.

We focus on the metrics because that's how our value is measured. That's how we demonstrate competence, secure resources, advance our careers, and provide for our families.

The salesperson concentrates on closing deals rather than building relationships. The professor prioritizes publication counts over meaningful research. The healthcare administrator focuses on bed turnover rather than healing environments.

Educational research has extensively documented that the phenomenon of "teaching to the test" has become increasingly tied to teacher evaluations, school funding, and administrative decisions. The *American Educational Research Journal*[26] shows that teachers spend significantly more time on test preparation than on creative or critical-thinking activities. Research from the National Education Association[27] found that in many districts, 60 to 80% of instructional time is devoted to tested subjects and test preparation strategies.

Academic studies have documented[28] how this narrowing of the curriculum correlates with decreased student engagement and creativity. Teachers report feeling forced to abandon proven practices in favor of test-focused instruction. The phenomenon has been termed "curriculum narrowing" by education researchers, who have documented how subjects like art, music, and even science and social studies receive less attention when they're not part of high-stakes testing regimes.

The use of patient satisfaction scores in healthcare has led to what medical researchers refer to as "the customer service model of medicine.[29]" Studies published in major medical journals[30] have documented how tying physician compensation and hospital funding to satisfaction scores can lead to medically inappropriate care. Research shows physicians may prescribe unnecessary antibiotics, order excessive tests, or avoid difficult but necessary conversations to maintain high scores.[31]

A study from UC Davis[32] found that patients who reported the highest satisfaction scores had higher healthcare costs, higher rates of hospitalization, and even higher mortality rates. Medical ethics publications report[33] how emergency department physicians face pressure to prescribe opioids and other controlled substances to maintain satisfaction scores. Medical ethicists have documented[34] how this system creates conflicts between optimal medical care and institutional pressures for high ratings.

I've caught myself in this pattern as well, measuring the success of phishing exercises by click-rate failures rather than the meaningful transformation of people reporting risks, and drafting content with an eye toward an arbitrary metric rather than genuine impact.

It starts as adaptive behavior, where we are simply responding to the incentives presented to us. But over time, it becomes something more: an internalized value system that gradually replaces our original purpose.

The human cost is twofold. First, we begin to view others through the system's lens, as metrics rather than people, as conversion opportunities rather than beings with inherent dignity. Second, we begin to evaluate ourselves by the same diminished standards.

Adapting Uncritically

When new technologies emerge, new metrics are introduced, or new algorithms determine what matters, how often do we pause to question them?

In my work with organizations implementing new systems, I've observed how rarely fundamental questions arise. Instead, the overwhelming response is uncritical adaptation. There is a collective assumption that technological change is inevitable, progress is linear, and our only choice is to keep pace or be left behind.

I've been guilty of this myself, adopting new platforms or processes without sufficient reflection, telling myself I'm "staying current" when I'm really surrendering critical judgment.

This uncritical adaptation means we often become carriers for system values we've never consciously evaluated or chosen. We implement changes that reshape work cultures, learning environments, and social dynamics without fully considering their human impact.

I've watched vibrant workplaces transform into metric-obsessed environments where innovation withers. I've seen educational institutions abandon curiosity-driven learning in favor of assessment-driven compliance. And in nearly every case, the change happened not through direct mandate but through the accumulated micro-decisions of people trying to adapt to what seemed inevitable.

Accepting Efficiency as the Highest Goal

Of all system values that we internalize, perhaps none is more seductive than efficiency.

Efficiency promises to save time, reduce effort, and maximize output. It whispers that if we just optimize thoroughly enough, we'll finally have space to breathe, to think, to be fully human.

But this promise contains a profound paradox: Pursued as an ultimate value, efficiency tends to eliminate the very things that make life meaningful.

Creative breakthroughs require inefficient exploration. Deep relationships depend on "unproductive" presence. Wisdom emerges from reflection that has no immediate payoff. Ethical clarity comes from contemplation that can't be measured.

I've worked with leaders who streamlined their organizations so effectively that they eliminated the spaces where innovation had previously flourished: the coffee break conversations, the experimental projects, the cross-disciplinary collaborations that didn't align with immediate deliverables.

We tell ourselves we're too busy to reflect, but the constant busyness itself represents the system's triumph. When efficiency becomes the highest value, meaning becomes a luxury we can never quite afford.

Without Resistance, We Volunteer for Our Chains

The uncomfortable truth at the heart of this section is this: Without conscious resistance, we willingly participate in the very systems that diminish us.

We reinforce the metrics by orienting our work around them. We validate the algorithms by adapting to their incentives. We perpetuate the values by modeling them for others.

And then we wonder why we feel so exhausted, so disconnected, so hollow.

The system doesn't need to force us into obedience. It only needs to make compliance feel like the only realistic path to survival and success.

Reclaiming Human Agency

We've journeyed through difficult terrain in this chapter: naming the ways systems can enslave us, acknowledging how we become complicit, and facing the psychological toll of diminished agency and moral fatigue.

But I can't end here, in the shadow of these realities, because my own experience has shown me that this is not where the story ends.

It's where the story takes a turn toward something new.

Machines Are Tools, Not Masters

We must remember something both ancient and urgent: We built the machines. We designed the systems. We programmed the algorithms. Now let that sink in for a moment.

They are extraordinary in their capabilities and astonishing in their complexity. But they are not sovereign.

The moment we forget this fundamental relationship, we surrender not just control but responsibility. We abandon our role as stewards of technology and become its subjects.

I've worked with tools that demanded both respect and caution. I've come to understand their power, potential, and risks. But I've also known this: A tool is never the source of purpose or vision. It's an extension of the hand that guides it and the mind that shapes its use.

Today's tools are exponentially more complex, but this fundamental relationship remains unchanged. Purpose, meaning, and wisdom reside in human hearts and minds, not in systems, no matter how sophisticated.

The Moment of Choice

Agency begins in the moment we reclaim choice, when we remember that between stimulus and response lies the power to choose our way.

Sometimes this choice appears in decisive moments, such as when we challenge a metric that distorts human value, when we design technology that enhances worth rather than diminishing it, or when we say no to an "improvement" that sacrifices meaning for efficiency.

I had a colleague who walked away from a prestigious position because the work he was doing was undermining the very values he cherished. It wasn't a dramatic exit. There was no manifesto, no public statement. Just a quiet, firm decision to align his work with his humanity. (He is thriving today because of that decision.)

That's agency. That's reclamation.

But more often, choice appears in small, daily moments that seem insignificant in isolation yet collectively shape the terrain of our lives: Turning off notifications during dinner with loved ones. Questioning a metric that doesn't capture what truly matters. Making space for unstructured thought and genuine connection.

These aren't revolutionary acts, but they are revolutionary in their cumulative effect, tiny reclamations of human space within increasingly mechanized environments.

Practices of Reclamation

If we are to reclaim our agency within systems designed to diminish it, we need practices that rebuild those muscles daily:

- Pause to reflect. Create regular spaces to ask yourself: What am I optimizing for? Is it aligned with what I truly value? What would change if I oriented around different metrics?
- Challenge the default. Question the necessity of each system that shapes your life. Does this need to be measured? Automated? Tracked? What would happen if we approached this differently?
- Rehumanize your work. Actively look for the person behind the profile, the humanity behind the metric. Build relationships that transcend transactional exchanges.
- Protect time for depth. Carve out and fiercely defend spaces for the kinds of thinking and connection that efficiency metrics will never value: reflection, exploration, and presence without agenda.
- Mentor with integrity. Show the next generation what it looks like to use technology without being used by it. Model what it means to lead with both technological fluency and human wisdom.

These practices aren't about rejecting technology or retreating from modern life to live in a cabin in the woods. They're about restoring technology to its proper role as a servant rather than a master, as a tool for human flourishing rather than a system that defines what flourishing means.

Become a Steward, not a Servant

The future will not be saved by those who worship at the altar of optimization or advancement for its own sake.

It will be saved by those who remember technology's place in the human story – as a powerful tool that must remain subordinate to human purpose, dignity, and wisdom.

We need stewards, not servants. We need people willing to stand at the intersection of technological power and human vulnerability and say: "I will not sacrifice soul for efficiency. I will not trade connection for convenience. I will not surrender moral agency for ease."

Closing Hope

I won't pretend this path is easy. Every day, I feel the pull of these systems myself — the subtle pressure to optimize, to perform for the algorithm, and to measure success by metrics that may have little to do with what truly matters.

However, here's what gives me hope: The fact that you've picked up this book and are reading these words means you've probably already sensed the tension. You've already heard that quiet voice within that whispers: *There must be more than this. We are more than what can be measured and optimized.*

That voice – persistent, inconvenient, refusing to be silenced – is the seed of your freedom.

So go ahead. Be the human the system didn't account for. Use the tools without being used by them. And remember that your value was never determined by your throughput.

It was always, and will always be, in your immeasurable, unoptimizable, and irreducibly human heart.

Chapter 5
Rebellion & The Radicalization Pipeline

The Moment of Turning

After my tenure in cybersecurity consulting at a Managed Service Provider, I made a deliberate career pivot into executive protection. This transition represented a fundamental shift in focus, from safeguarding digital assets to protecting physical ones: people and their families.

During this period, I served as an instructor for the company's executive protection training program, where I specialized as an adversarial role-player in Field Training Exercises (FTXs). My responsibility was to simulate the behaviors and strategies of potential attackers, essentially conducting what we would call "Red Team" operations in cybersecurity.

This immersive experience offered profound insights into the psychology and behavioral patterns of attackers. I developed a reputation among professional protectors as a particularly formidable and unpredictable adversary, though those specific experiences merit a discussion for another time.

Following this work, I conducted extensive research on assassination attempts, attacks on high-value individuals, and what we once called "active shooters" (now more comprehensively termed "active assailants" or "mass shooters").

This research culminated in a collaboration with Grant Cunningham, a distinguished author and instructor specializing in personal security and defensive shooting. Together, we wrote *Praying Safe: The Professional Approach to Protecting Faith Communities,* a comprehensive examination of how worship communities can protect themselves against radicalized attackers.

My background afforded me a specialized perspective on attacker psychology, which, when combined with Grant's expertise and guidance, resulted in a book that resonated strongly with worship communities.

I believe this experience led me to a critical realization: The technology sector I originally came from is cultivating some of the same patterns of radicalization we identified in worship communities. There is a concerning shift emerging among certain individuals that, as a security professional, I find deeply troubling. For those working in security, particularly in executive protection, I urge heightened awareness. We are witnessing a transformation in how people respond to the transitions described throughout this book.

Our world today has perfected the conditions for this transformation.

We live in an era of unprecedented connection, where billions of people carry devices that can reach almost anyone, anywhere. Yet, loneliness has become an epidemic, spreading through our societies like a silent contagion.

We have ready access to more knowledge than any previous generation, with search engines, digital libraries, and instant information at our fingertips. Yet, many still feel profoundly powerless to change anything that truly matters in our lives or communities.

Our tools offer convenience beyond what our grandparents could have imagined: food delivered with a tap, entertainment streaming endlessly, work that follows us everywhere. But these tools simultaneously strip away the dignity that comes from meaningful labor, genuine connection, and true agency.

The result is a quiet, persistent erosion: an erosion of belonging, an erosion of meaning, and an erosion of the fragile belief that life inside the increasingly automated machine is worth the emotional cost of being a part of it.

This erosion doesn't always create dramatic rebels who make manifestos and take to the streets. At least, not at first. Instead, it creates a subtle withdrawal, a pulling back, and a quiet questioning that goes unanswered.

It's in this silence, in the empty spaces where meaning should live, that the seeds of radicalism are sown — not in grand speeches or radical philosophies, but in the silent rooms where people sit alone, wondering if the world has forgotten them.

Many will bury this emptiness in distraction, the endless scroll, the next show to binge, the next purchase promising momentary fulfillment. But a dangerous few will begin to ask the old, terrible questions: "Why should I stay inside a system that no longer sees me as human? Why not make it hurt the way I hurt?"

This reminds me of the commonly attributed African proverb, "The child who is not embraced by the village will burn it down to feel its warmth."

When it forgets the human beings within it, the system doesn't merely create disconnection. It creates risk. It plants the seeds of its own backlash.

The story of rebellion doesn't begin with violence in the streets. It begins with silence in the soul.

A Wave of Modern Rebels

I remember the 1990s as a decade when extremism seemed to erupt from every corner of American society. I watched news reports of the Oklahoma City bombing in stunned disbelief — 168 people, including children in a daycare center, killed in what remains the deadliest act of domestic terrorism in U.S. history. Just a year later came the bombing at Atlanta's Centennial Olympic Park during the 1996 Summer Olympics, followed by the horrific Columbine High School shooting in 1999 that seemed to signal a new era of mass violence.

These events didn't occur in isolation. They followed the traumatic government sieges at Ruby Ridge in 1992 and Waco in 1993, confrontations that became rallying cries for anti-government extremists across the country. Each incident further eroded trust between segments of the American public and their government, creating fertile ground for conspiracy theories and extremist ideologies.

Looking back now, what's striking is how this wave of extremism coincided with America's rapid technological and economic transformation. The 1980s and early 1990s witnessed a decline in manufacturing employment as automation transformed factories across the Rust Belt. Between 1979 and 1983 alone, manufacturing lost 2.4 million jobs, nearly 12% of its workforce, with steel towns in Pennsylvania, auto manufacturing centers in Michigan, and countless other industrial communities devastated almost overnight.

These weren't isolated economic statistics — they represented real communities torn apart. Towns like Youngstown, Ohio lost 50,000 jobs in the steel and related industries, while Flint, Michigan witnessed General Motors reduce its workforce from 80,000 to under 8,000 over a two-decade period. Many of these communities have yet to recover. What emerged in this vacuum wasn't just unemployment, but profound grievance, alienation, and a search for meaning.

When we examine the extremist movements that flourished during this period, disturbing patterns emerge. The anti-government militia movement that inspired Timothy McVeigh and Terry Nichols grew in direct response to economic dislocation and perceived government overreach. The Army of God targeted abortion providers, finding moral justification for violence in religious extremism. Eric Rudolph, the Olympic Park bomber, combined anti-government sentiment with religious fundamentalism in his campaign against abortion clinics and a lesbian nightclub.

Meanwhile, Theodore Kaczynski, the Unabomber, waged his own one-man war against technological society from an isolated cabin in Montana. His manifesto articulated a radical critique of industrial civilization that resonated with many who would never have condoned his methods. "The Industrial Revolution and its consequences have been a disaster for the human race," he wrote, expressing an alienation from technology that found twisted expression in violence.

These domestic movements were paralleled internationally by groups such as the Irish Republican Army in Northern Ireland, the Red Army Faction in Germany, and Aum Shinrikyo in Japan, each responding to different grievances but all sharing a profound alienation from mainstream society and a belief that their cause justified the use of violence.

What's crucial to understand is that these weren't simply isolated "madmen" or "monsters." Each of these movements emerged from recognizable patterns of human response to alienation and displacement:

- Grievance created anger and justification for revenge. For McVeigh, it was Waco and Ruby Ridge; for the IRA, it was centuries of British rule.
- Economic dispossession added resentment and fear of decline. The loss of traditional manufacturing jobs and rural ways of life fed into militia movements that offered an explanation and someone to blame. The loss of meaning created a vacuum that radical purpose filled. Aum Shinrikyo offered apocalyptic meaning to educated Japanese youth disillusioned with material success.
- The Columbine shooters found meaning in a narrative of revenge against a society they felt had rejected them.

- Technical alienation drove anti-system, anti-modern sentiments. Kaczynski's rejection of the industrial-technological system represented an extreme version of a broader discontent with the accelerating degradation of natural human life. Isolation weakens social ties and heightens vulnerability to extreme ideas. The cabin in Montana created a perfect storm of physical isolation that cultivated his ideological extremism.

These patterns aren't confined to history. As AI threatens to transform work and society at a pace that dwarfs previous technological revolutions, these same factors are converging with unprecedented force. The warehouse worker replaced by robots, the truck driver facing obsolescence from self-driving vehicles, the middle manager whose job is streamlined by algorithms — all experience the same psychological pressures that have historically driven extremism.

We cannot afford to look away from this reality, because when we fail to acknowledge genuine human suffering, when we dismiss the alienated and disaffected as simply "disturbed," "maladjusted," or "dangerous," we do not eliminate the risk they pose.

We amplify it.

A society that systematically ignores the humans being displaced, discarded, and devalued by its pursuit of technological progress is a society that is inadvertently cultivating its own opposition.

The lesson of the 1980s and 1990s is not that extremists were uniquely evil during that era, but that predictable conditions produced predictable results. When people lose economic stability, they lose meaningful work, social connection, and a sense of purpose. They feel the systems around them are indifferent to their suffering; a percentage will always turn to extremism.

We need to face this reality squarely, not to excuse violence or extremism, but to understand its roots deeply enough to address them before they bear bitter fruit. By recognizing the pattern of factors that consistently lead to extremism, from grievance to isolation to moral justification, we can work to mitigate these conditions before they give rise to the next wave of modern rebels.

Conditions That Breed Extremism

Throughout my years studying the psychology of radicalization, I've come to understand that rebellion doesn't erupt spontaneously. It grows gradually, taking root in the cracks and fissures of a society that has lost sight of the human beings within it. It has a clear and distinct runway.

If we want to understand why some individuals move from frustration to fury, from alienation to action, we need to honestly examine the soil in which these transformations are planted.

Let me share what I've learned about the conditions that breed extremism, not as abstract theories, but as lived realities I've witnessed in my work and study.

Economic Dispossession

Economic dispossession isn't merely about financial hardship. It's also about the profound sense of being pushed to the margins, of being rendered irrelevant, disposable, and forgotten.

We observe this pattern recurring across various industries and regions. Factory workers watch as jobs continue to disappear due to automation. Service workers are seeing their hours and benefits cut by algorithmic scheduling. Middle-aged professionals are being laid off and told to "re-skill" for a job market that seems more and more indifferent to experience.

Initially, this dispossession creates despair, a quiet resignation that can look deceptively like acceptance. But if left unaddressed, that despair often evolves into resentment. And if that resentment finds no constructive outlet, no path toward meaning and reintegration, it can metastasize into radicalization.

History has shown this pattern repeatedly. The most dangerous social movements often take root not in conditions of absolute poverty, but in the volatile mixture of humiliation, loss of status, and shattered expectations.

When people believe the system has betrayed them, that the social contract promising dignity in exchange for participation has been broken, some will eventually decide to break it themselves.

Loss of Meaning

In the endless pursuit of efficiency, optimization, and growth, our systems have often neglected something essential: the human need for meaning.

When life inside the machine feels hollow, when work feels disconnected from purpose, when community feels fragmented or superficial, and when achievement feels empty, people naturally go searching for something more.

They search for purpose, identity, belonging. And if the mainstream system offers no meaningful answer to these deep human hungers, other voices will step in to fill the void.

Extremist ideologies don't recruit with rational arguments. They recruit with promises of purpose.

They offer clear villains to explain complex problems. They provide noble missions to replace empty careers. They transform the isolated and powerless into warriors, the invisible into symbols, and the lost into believers with a cause larger than themselves.

The lesson is stark but clear: If we drain ordinary life of meaning and purpose, we leave people vulnerable to dangerous alternatives that promise to restore what has been lost, often at a terrible price.

Technological Alienation

There is a unique form of rage that emerges from feeling controlled by forces you cannot touch, see, or influence.

When a digital platform mediates every transaction, when every inter-action is shaped by algorithms, and when every decision is reduced to data points, something vital in the human spirit begins to crack.

Most people affected by this technological alienation don't turn toward violence. They adapt, compartmentalize, and find ways to carve out meaning in the margins of a system they cannot change.

But a dangerous few choose another path: They decide to fight back against a system they experience as fundamentally hostile to their humanity.

We're already seeing early warning signs. We have seen attacks on autonomous vehicles, sabotage of smart city infrastructure, vandalism targeting symbols of technological control, and isolated violence against tech leaders and companies.

These are not random acts of destruction or simple criminal behavior. They are symptoms of a deeper, spreading anger at systems that seem to advance at the expense of human dignity and agency.

The case of the Unabomber represents an extreme example of techno-logical alienation combined with isolation and loss of meaning. His manifesto reveals a mind that perceived modern technology as funda-mentally incompatible with human freedom and dignity, driving him to violence as a form of resistance.

Grievance and Perceived Injustice

Perhaps the most powerful fuel for extremism is grievance, a deep, burning sense of having been wronged, either personally or as part of a group identity.

Grievances can come in many forms. There are historical wrongs that have never been acknowledged or addressed. Systemic discrimination, real or perceived, is fuel for outrage. Some organizations tell people that their cultural or religious identity is under attack in order to create anger artificially.

What makes grievance particularly dangerous is its self-reinforcing nature. Once someone begins to interpret the world through the lens of victimhood, nearly every subsequent experience can be filtered to confirm and intensify that perspective.

We see this clearly in movements like the IRA, where historical grievances over British rule in Northern Ireland created a multi-generational cycle of violence, with each new incident serving to justify further retaliation.

The danger deepens when grievances are collectivized and personal wounds become merged with group narratives of suffering. Individual pain, when validated and amplified by a community that shares similar experiences, transforms into something more potent: a righteous cause demanding action.

Identity and Belonging

Human beings are social creatures at our core. We need to belong and to feel welcomed, valued, and understood by others. When mainstream society fails to provide this essential connection, extremist groups often step into the void with a powerful alternative.

Extremist groups excel at providing what psychologists call a "total identity," a complete framework that answers the fundamental questions of human existence:

When someone asks, "Who am I?" these groups provide a ready answer: You are a warrior, a defender, a true believer. When the searching soul wonders, "What matters in this world?" the response is clear and straightforward: Our cause matters, our people matter, and our struggle defines everything. To the question, "Where do I belong?" they offer the stark binary of absolute belonging: You belong with us, standing against them. And perhaps most powerfully, when someone asks that very human question, "What gives my life meaning?" these groups provide a purpose that feels both urgent and transcendent: Your meaning comes from fighting for our survival.

These groups create powerful bonding rituals that forge deep emotional connections among members. They establish clear boundaries between insiders and outsiders. They provide a ready-made community with shared language, symbols, and purpose.

We have seen this pattern in groups from Al-Qaeda to organized crime gangs, where the sense of brotherhood and shared mission creates bonds stronger than family ties. The more isolated or alienated from mainstream society someone feels, the more powerful this offer of belonging becomes.

What makes this particularly dangerous is that once the bond of identity is formed, leaving the group becomes almost unthinkable, not just because of potential retaliation, but because it means losing one's entire sense of self. The individual's identity becomes so fused with the group that separation feels like a form of death.

Binary Thinking

Complex problems rarely have simple solutions. Yet extremism thrives on certainty, on dividing the world into absolute categories of right and wrong, us and them.

This binary thinking creates critical conditions for extremism: It simplifies a complex world into manageable categories. It eliminates the discomfort of moral ambiguity. It creates clear enemies responsible for all suffering. It justifies extreme measures against those enemies.

When society seems overwhelmingly complex and chaotic, the comfort of absolute certainty becomes almost irresistible. The messy, contradictory nature of reality is replaced with a narrative that explains everything through a single, coherent framework.

This is why we see such strong correlations between extremism and authoritarian personality traits. Individuals who crave certainty and clear moral boundaries are particularly vulnerable to ideologies that promise exactly that.

The digital age has intensified this tendency, as algorithms drive people toward progressively more extreme content and social media creates isolated bubbles where simplistic worldviews can flourish unchallenged by opposing perspectives.

Moral Justification

For most people, violence requires overcoming powerful moral inhibitions. Extremist ideologies provide frameworks that transform harmful acts from moral violations into moral imperatives.

This moral inversion operates through several mechanisms: Dehumanization portrays opponents as less than fully human. Moral disengagement creates psychological distance between actions and consequences. A transcendent purpose frames violence as necessary for the greater good. Sacred duty transforms violence from choice into obligation.

When someone believes they are acting in accordance with divine will, historical necessity, or cosmic justice, almost any action can be justified. The moral constraints that normally govern human behavior are not simply ignored — they are systematically inverted.

We see this clearly in groups like Aum Shinrikyo, whose members believed their violence was actually compassionate, releasing victims from a corrupt world. Similarly, eco-extremists sometimes frame their actions as necessary sacrifices to save countless future lives threatened by environmental collapse.

The power of moral justification explains why many extremists exhibit little to no remorse for their actions. In their minds, they haven't broken moral laws; they've upheld them at the highest and most courageous level.

Echo Chambers and Radicalization Environments

Ideas don't radicalize in isolation. They intensify through reciprocal reinforcement within closed information ecosystems where extreme views are constantly validated, and opposing perspectives are systematically excluded.

These echo chambers operate through several mechanisms, including selective exposure to information that confirms existing beliefs: social reinforcement from like-minded others, gradual normalization of increasingly extreme positions, and devaluation of outside sources as corrupt, biased, or uninformed.

The digital landscape has vastly accelerated this process. Algorithms optimize for engagement, which often means feeding users extreme content. Online communities enable people with fringe beliefs to connect across vast distances. Personalized news environments create the illusion that one's worldview is more widely shared than it actually is.

We saw this pattern clearly in Timothy McVeigh's radicalization, where his immersion in gun shows, extremist literature, and isolated compounds gradually shifted his perception of reality until violent resistance seemed not just justified but necessary.

What makes echo chambers particularly dangerous is their invisibility to those inside them. Fish don't know they're in water, and people embedded in ideological bubbles rarely recognize the boundaries of their information environment.

The Combined Effect

When we combine these conditions — economic dispossession, loss of meaning, technological alienation, grievance, identity needs, binary thinking, moral justification, and echo chambers —we create an environment where rebellion becomes not just possible but, for some, inevitable.

At the core of every significant rebellion throughout history lies a simple, terrible truth: Someone believed they had no place left to stand. No voice that would be heard. No future worth enduring.

When human worth is systematically erased, and people are reduced to productivity metrics or data points, we create the perfect breeding ground for ideologies of destruction.

This is not offered as a defense of those who choose violence. It is an indictment of the conditions that make violence appear to some as the only language left to speak.

We cannot fight rebellion effectively with force alone. We must fight it by restoring the dignity, meaning, and agency that its absence hollowed out.

Understanding the New Rebels

During my research on technological resistance movements, I've come to recognize that our cultural imagination of rebellion is often too narrow and too simplistic to capture the complex reality unfolding around us.

When most people hear the word "rebel," they picture dramatic protests, perhaps even violence, or an overt rejection of technology or society. But today's landscape of resistance is far more nuanced, diverse, and often invisible to casual observation.

We need to understand this landscape if we hope to distinguish between healthy dissent, which can strengthen our society, and destructive rebellion that threatens to tear it apart.

Not All Rebels Are Anti-Technology

The perception that those who challenge technological systems are simply anti-progress is far from the truth. At gatherings like DefCon, Black Hat, and BSides — the premier conferences where the hacker community congregates — one finds not technological Luddites, but some of the most technically sophisticated minds in the world.

These communities represent a crucial distinction that is often missed in discussions about technological resistance. Many of today's rebels are not rejecting technology itself; they are rejecting its misuse. They are fighting for a different relationship with it. They want technology that respects human boundaries, operates with consent and transparency, and augments human capability without diminishing human agency.

At these conferences, hackers demonstrate vulnerabilities in everything from voting machines to medical devices to smart city infrastructure. They do this not to undermine these systems, but to strengthen them, ensuring that the technologies we depend on more and more are worthy of our trust.

"Hacking" in its purest form is about understanding systems deeply enough to make them serve human needs better. It's about questioning assumptions, finding creative solutions, and maintaining human control over the tools we create. The hacker ethos is fundamentally about empowerment, not destruction.

These digital rebels hack systems not to destroy them, but to expose their flaws and limitations. They build encrypted networks not to escape society, but to create spaces within it where genuine freedom remains possible. They challenge automation not out of nostalgia, but out of concern for a future where human judgment becomes obsolete.

Their rebellion is not against progress. It's against a particular vision of progress that seems to have forgotten why we build these systems in the first place: to serve human flourishing rather than replace it. They insist that technology should remain a tool in human hands, not the other way around.

In this sense, these rebels represent not a rejection of our technological future, but a vital force that ensures the future remains fundamentally human.

Invisible Rebellion

Not all rebellion takes place in visible spaces.

In my work documenting resistance movements for my previous book, I encountered forms of rebellion that operate beneath the surface of everyday awareness: Developers inside major tech companies are quietly subverting harmful projects from within. Communities are developing alternative economic systems that operate outside mainstream financial structures. Operating beyond the reach of surveillance, individuals are creating tools that enhance privacy and autonomy. Open-source encrypted messaging platforms that have seen explosive adoption whenever surveillance concerns spike, provide secure communication channels outside traditional monitoring systems.

This invisible rebellion operates through code, networks, and ideas that move beneath the surface of society, like underground rivers, reshaping the landscape in ways that may not be immediately apparent.

We live in an era when power is becoming invisible, operating through algorithms, financial systems, and information architectures that most citizens never see or understand. In response, rebellion has learned to match this invisibility, working in the shadows where power now resides.

These hidden forms of resistance are often more significant than their visible counterparts, creating alternative possibilities that flourish in the spaces among our existing systems.

The Spectrum of Resistance

Most resistance to dehumanizing systems is non-violent. It comes in the form of whistleblowers, journalists, ethical technologists, artists, organizers, and everyday dissenters. It challenges the system not with destruction, but with questions. Not with force, but with alternative visions.

We need to recognize and protect this space for healthy dissent, for the voices that challenge us to build better systems, and to remember the humans these systems are meant to serve.

History offers a sobering lesson: When peaceful resistance is ignored, dismissed, or suppressed, it creates conditions that make violent outliers more likely to emerge.

A system's refusal to listen to constructive criticism doesn't make that criticism disappear. It merely ensures that the criticism will eventually take forms that cannot be ignored.

This is not just a moral observation. It is a practical one. The path to a stable, flourishing technological society runs through our ability to listen to its critics, not despite them.

Healthy Dissent vs. Destructive Rebellion

The challenge for any society navigating rapid technological change is to recognize the difference between dissent that strengthens us and rebellion that threatens to tear us apart.

Healthy dissent asks us to live up to our stated values. Destructive rebellion rejects the possibility that those values were ever genuine.

Healthy dissent challenges specific practices while accepting shared principles. Destructive rebellion rejects the legitimacy of the entire system.

Healthy dissent seeks to reform. Destructive rebellion seeks to remove.

We need to create spaces where healthy dissent can thrive, uncomfortable questions can be asked, alternative approaches can be explored, and the human costs of progress can be honestly acknowledged.

Because when those spaces collapse and people believe they will never be heard, the most dangerous actors move from the margins to the center of resistance.

Warning Signs of Brewing Rebellion

Throughout history, major societal upheavals have rarely arrived without warning. . Most people sensed a clear uneasiness growing but underestimated its danger until it was too late.

We are already living among the warning signs of a brewing rebellion against dehumanizing systems. Still, we too often dismiss these signs as fringe phenomena, outliers, or noise rather than recognizing them as signals.

We cannot afford that willful blindness any longer.

Let me share what I've observed about the early indicators that should command our attention.

Online Radicalization Communities

The first warning signs appear not in physical spaces but in digital ones, in the corners of the internet most of us never visit.

Across the digital landscape, a concerning pattern has emerged in online communities focused on technological displacement. What often begins as support forums where people share legitimate concerns about automation and job loss can evolve over time. These spaces transform from places of mutual support into environments where more radical perspectives take root, sometimes culminating in discussions of direct action against technological systems.

This transformation isn't merely speculative. We've seen similar patterns of online radicalization across various contexts, from political extremism to anti-social movements. The progression follows a recognizable path as participants move from articulating grievances to forming collective identities based on shared experiences of displacement, then to shared intent:

Forums where the economically displaced gather to process their loss. Channels where those who feel betrayed by technological "progress" seek understanding. Communities where alienation from mainstream values cultivates alternative moral frameworks.

What makes these spaces particularly dangerous is not just their content but the human hunger they reveal – the hunger to be seen, to have one's pain acknowledged, and to matter in a world that seems indifferent.

When these legitimate human needs find no healthy outlet, they don't disappear. They transform, hardening from pain into rage, and from isolation into a collective identity built around rejection of the system perceived as the source of suffering.

Physical Manifestations

Digital discontent can sometimes manifest in the physical world, revealing deeper tensions about the role of technology in society.

While not widely covered in national media, there have been documented instances of resistance to physical technological systems in various communities. I have read news reports of vandalism to automated systems in some urban areas and concerns about surveillance technology leading to protests or resistance. There have been controversies surrounding the expansion of infrastructure, such as cell towers or data centers, in communities all over the world.

These incidents, when viewed collectively rather than as isolated events, may represent points where digital anxiety transforms into physical action.

These responses are often dismissed as extreme or irrational, but they can signal legitimate concerns about privacy, autonomy, and the pace of technological change. As such, they deserve thoughtful consideration.

Emerging Ideological Frameworks

Perhaps the most significant warning sign is the emergence of coherent ideological frameworks that provide meaning and direction to technological resistance.

Several emerging movements have begun to challenge the dominant narrative of technological progress, each offering alternative perspectives on our relationship with technology. Some advocate for "digital minimalism," focusing on intentional and limited use of technology. Others promote "right to repair" initiatives, pushing back against planned obsolescence and the practice of designing products to be disposable. Communities exploring "tech sabbaths" and digital detox practices are growing in various parts of the world.

These movements don't necessarily reject technology outright but instead question our current trajectory and advocate for more human-centered approaches to innovation.

While these perspectives remain outside the mainstream discourse, they represent important counterpoints to uncritical technological adoption, raising valid questions about who benefits from rapid digital transformation.

When these ideologies intersect with communities experiencing acute alienation, they can transform diffuse discontent into focused action by providing explanatory narratives that make sense of personal suffering, moral frameworks that justify resistance, and strategic visions that direct energy toward specific targets.

Whether these movements evolve toward constructive reform or destructive rebellion depends largely on how mainstream institutions respond to their emergence.

Systemic Deafness Breeds Escalation

Across all these warning signs runs a common thread: When legitimate grievances go unacknowledged, resistance escalates.

The most dangerous thing we can do right now is dismiss the pain driving people toward resistance.

When society refuses to listen to human suffering and treats alienation as a personal failing rather than a collective responsibility, it unwittingly pushes people toward more extreme forms of expression.

We need to listen — not to excuse destructive action and not to abandon progress, but to understand the human costs of our current direction, because understanding is the only foundation for prevention.

Preventing Violent Rebellion

I've come to believe that we are not powerless in the face of brewing rebellion. We are not condemned to cycles of progress and backlash, innovation and destruction.

However, addressing this challenge requires that we understand what is truly at stake.

Beyond the False Choice

In the aftermath of technological resistance movements, society often falls into a familiar pattern: We amplify security, criminalize dissent, and double down on the very approaches that sparked resistance in the first place, just as British authorities did with the Luddites in the 19th century when they responded to worker fears with force instead of understanding.

We tell ourselves that the only answer to rebellion is to crush it, that the march of progress must continue uninterrupted, that those who question it are simply obstacles to be overcome.

But this reaction misses the fundamental issues entirely.

The problem isn't the technology itself. The problem is how we design, deploy, and govern it. It's also not the dissent. The real problem is our systemic failure to listen to each other before we harden ourselves into rejection.

This blindness to the human dimension isn't just a moral failure. It's a practical one. It leads us to build systems that generate opposition.

We don't have to choose between innovation and human welfare. We don't have to sacrifice meaning on the altar of efficiency. These are false dichotomies that ultimately serve no one.

The Path of Human-Centered Renewal

What would it look like to build a world that honors human dignity, meaning, and agency alongside technological advancement?

I've had the privilege of witnessing glimpses of this alternative path in various settings while working as a security consultant. I was exposed to various companies across nearly every sector. One of those companies stands out as being far superior to the others. The owners and executive leadership of the company understood that if their employees were happy, they would pass that along to their customers. And they did. The customers could feel the positive energy and see people going out of their way to help them. The company was not just measuring success by productivity metrics, but also by employee well-being and sense of purpose.

This approach separated them from their competitors. The level of customer service and care the employees gave was a market differentiator. They were a force to be reckoned with in their industry. Oh, and technology was at the forefront of what they were doing. In fact, the employees were the ones finding ways to use technology efficiently. Their competitors were left scratching their heads, trying to figure out how they were falling behind when they were using the same technology.

Examples like this share a common approach: They treat technology as a means rather than an end, as a tool to enhance human capability rather than a replacement for human judgment.

The path forward requires that we widen our definition of progress beyond narrow metrics of efficiency and economic growth.

True progress isn't just faster machines or higher profits. It's deeper humanity, richer connections, more meaningful work, and the power to shape our collective future.

The Ethical Imperative

If we continue on our current trajectory, accelerating technological development without equal attention to its human impact, dismissing early warning signs of alienation and resistance, and treating those who question our direction as obstacles rather than essential voices, we are not just risking rebellion.

We are guaranteeing it.

But if we build systems that remember the humans within them, design technologies that enhance agency rather than diminish it, and create economies that value contribution beyond narrow productivity, we do more than avoid potential catastrophe.

We create the conditions for genuinely sustainable progress that carries everyone forward.

Rebellion becomes inevitable when care disappears from the system. Renewal becomes possible when care returns to the center.

A Personal Commitment

Here is the charge I leave you with, as a leader, a citizen, and a human being:

Listen now. Care now. Act now.

Do not wait for the system to break. Do not wait for the rebels to rise.

The future we are creating is not merely a technical project. It is a human one.

And if we get that right — if we remember the people inside the technology and systems we build — we may create a world where rebellion becomes unnecessary because humanity has become undeniable.

Chapter 6
Psychological Collapse

When Autonomy Disappears

The human body possesses a remarkable resilience that the mind cannot always match. This truth manifests in the quietest moments: in the distant gaze of colleagues during meetings, in the gradual hollowness that creeps into once-vibrant voices, in the scatter of our own thoughts when notifications pierce moments of focus.

Our bodies continue functioning. We wake, we work, we scroll, we sleep, even as something essential within us begins to crack under pressures we weren't designed to bear. While our physical forms adapt to technological environments, our cognitive architecture strains under the constant demands of digital life.

This cognitive strain becomes clearer when we consider the dual-process model[35] of thinking that psychologists have identified. Our "System 1" thinking — fast, intuitive, and automatic — evolved for a world of immediate physical threats and opportunities. Our "System 2," which is slower, deliberate, and analytical, developed to handle complex problems that require sustained attention.

Modern technology environments systematically hijack our System 1 through endless notifications and algorithmic feeds designed to capture attention, while simultaneously depleting the limited cognitive resources our System 2 needs for meaningful thought and decision-making.

The result is a profound disconnection from our autonomy, a sense that we're constantly reacting rather than choosing, consuming rather than creating, surviving rather than flourishing. This is the reality we must confront: The body can survive confinement longer than the spirit can endure purposelessness.

In the wealthiest, most technologically advanced societies humanity has ever known, our minds are fracturing at alarming rates. Depression and anxiety have reached unprecedented levels. Suicide rates climb steadily. Loneliness has become so prevalent that health authorities classify it as an epidemic. These aren't isolated problems but symptoms of a collective condition.

We've engineered a world optimized for efficiency, scale, and constant productivity. We've built systems that move at inhuman speeds, that never tire, never require stillness, and never need connection. And then we've demanded that humans adapt to these systems rather than designing systems that honor human cognitive limits and psychological needs.

The result isn't just burnout; it's a profound collapse of meaning, purpose, and connection. This collapse affects people across professions and generations, from creative fields to knowledge work to manufacturing, as technological systems increasingly dictate the rhythms and relationships of our lives.

It doesn't take a scientific study to observe this collapse around us, though research confirms[36] what many experience firsthand. Look at the quiet resignation in workplaces where people once found purpose, the growing difficulty many report in sustaining attention on anything meaningful, and the paradoxical isolation amid unprecedented connectivity.

Conversations with friends, colleagues, and even strangers often reveal a similar theme: a sense of being perpetually distracted, chronically exhausted, and somehow less present in our own lives. As our phones become more advanced, our attention spans grow shorter. As our technologies advance, our capacity for deep thought, meaningful connection, and sustained joy seems to be in retreat. This reflects a fundamental misalignment between what human minds need and what our technological environments provide.

This psychological collapse isn't a side effect of progress. It's the predictable outcome of a fundamental error: We've built a world that optimizes for everything except human well-being.

Naming the Wounds

Understanding this collapse requires us to shift our focus. We must stop blaming individuals for their inability to adapt to inhuman conditions and instead examine the environment we've created.

Corporate wellness programs and productivity seminars consistently frame psychological distress as an individual failure requiring personal solutions. This approach is evident in the language used across workplaces, healthcare settings, and even educational institutions. The emphasis remains on teaching people to adapt rather than questioning the systems that necessitate such adaptation.

We've developed a cultural habit of personal blame: If you're exhausted, it's likely that you lack proper boundaries. If you're anxious, you need better coping skills. If you're lonely despite being constantly connected, you must be socially deficient. If you're struggling to find meaning in your work, you're in the wrong career.

But what if we're asking the wrong questions? What if the wound isn't within the individual but in the environment we've constructed?

Studies in occupational health consistently demonstrate that workplace factors, including organizational culture, workload demands, and technological monitoring are stronger predictors of burnout than individual psychological traits. Research on digital well-being similarly shows that interface design choices directly impact user stress levels and attention capacity regardless of personal resilience factors.

The data points to a profound truth: We've normalized conditions that are fundamentally at odds with human well-being, then blame individuals for the natural human response to these conditions.

Before we can heal this fracture, we must name it honestly. We must recognize that the rising tide of depression, anxiety, disconnection, and despair isn't a collection of individual failings but evidence of systems that have forgotten what humans need to thrive.

The Psychological Shift

The transition from tool user to system component happens gradually, in a way so subtle that we may not even notice it.

First, we leverage technology to enhance efficiency. Then we optimize ourselves to meet algorithmic requirements. Performance metrics influence personal judgment almost to the point of replacing it. Engagement scores substitute for authentic connection. Productivity measures override meaningful work.

The psychological toll manifests itself in unexpected ways: Professionals who once took pride in expertise now question their relevance. Young people entering careers wonder if their chosen fields will exist in a decade. Seasoned workers watch institutional knowledge become commodified data.

This isn't just workplace anxiety. It's a crisis of human relevance. When machines excel at tasks once considered uniquely human, we face a fundamental identity crisis. If intelligence, creativity, and even emotional response can be automated, what anchors human worth?

The answer shapes whether technology enhances or erodes our humanity. How we respond to this displacement will determine whether we emerge stronger or find ourselves invisible within our own systems.

Our Emotional Response to Machines

Our relationship with technology has evolved from one of wonder to one of wariness. The early internet felt like discovering new frontiers. It was vast, open, and full of possibility. Social media initially connected us across distances, reuniting lost friends and creating communities around shared interests.

Now that wonder has soured into anxiety. Every notification demands an immediate response. Algorithms know us better than we know ourselves, predicting our desires while heightening and manipulating our fears. The same platforms that promised connection have become vectors for loneliness, comparison, and rage.

We've become paradoxical beings: hyper-connected yet profoundly isolated, constantly informed yet increasingly confused, endlessly entertained yet spiritually empty. The emotional toll is measurable in rising rates of depression, anxiety, and what some researchers call "digital fatigue," a state in which technological interaction becomes draining rather than enriching.

This emotional shift reflects a deeper psychological crisis. When our tools become our masters, when our devices demand more attention than our relationships, and when algorithms more than personal experience shape our worldview, we lose something essentially human: the capacity for authentic choice and genuine presence.

Constant Digital Mediation

Digital mediation has fundamentally altered our experience of life. The constant flow of notifications and the habitual checking of devices have become embedded in daily existence, and we often sacrifice moments of genuine connection for digital engagement. This pattern occurs so frequently that it becomes nearly invisible — a background process in our consciousness.

The cost of this constant mediation extends beyond mere distraction. It creates a profound shift in our relationship with reality itself. Through digital interfaces, we lose the grounding of physical senses — the irreplaceable presence of another person's energy, the natural rhythm of uninterrupted conversation, and the ability to be fully present without the compulsion to document or broadcast moments.

This digital mediation becomes even more concerning when we consider the rise of AI-generated content that blurs the lines between authentic and synthetic reality. The proliferation of deepfakes and AI-generated media has created what psychologists refer to as a "climate of uncertainty and distrust" for average users, who struggle to determine whether a video, image, or voice is real or AI generated without specialized forensic analysis.

While most of us experience this as a vague unease, for individuals with psychotic disorders, this uncertainty transforms into devastating delusions that reinforce paranoid worldviews. Their experience represents an amplified version of the reality blurring that affects us all to varying degrees. For some, the experience causes mild confusion about whether an image is AI generated or not.

Others experience a complete dissolution of the boundary between reality and fiction. This extreme example illuminates a broader truth: When our fundamental ability to trust sensory information erodes, so does our psychological stability. The question becomes not just whether we're present in our lives, but whether what we're present with is even real.

This division of attention creates a persistent disconnection from embodied reality. Consciousness becomes split between physical presence and digital space, transforming people from participants in their own lives to spectators. The mind remains partially elsewhere, creating a subtle but significant barrier to direct experience.

Productivity Obsession

The cultural fixation on productivity has evolved from a means to an end into an identity, a value system, and for many, a religion. What once might have been measured in moments of connection, insights gained, or depth of engagement is now quantified in outputs: emails answered, tasks completed, and projects delivered.

The initial sensation of productivity can feel exhilarating — the dopamine response from checking items off a list and the satisfaction of visible accomplishment provide temporary validation. Yet beneath this surface satisfaction often lies a deeper emptiness.

This productivity-obsessed paradigm takes a toll. The body continues functioning — waking earlier, working longer, optimizing every moment — while the mind and spirit erode under the relentless pressure to produce rather than to be. Performance metrics may improve while meaning and purpose diminish, creating a hollowness that no amount of efficiency can fill.

Identity Crisis and Disposability

Work has always been a central aspect of human identity. We define ourselves through our professions, derive meaning from our contributions, and build communities around shared vocational experiences. We were defined by the trades that shaped us, identities like Smith, Wright, Weaver, Cooper, and Farmer. Automation threatens this foundational aspect of human self-understanding, creating perhaps the most profound wound inflicted by current technological systems: the growing sense of human disposability.

When a radiologist watches AI diagnose scans with greater accuracy, a journalist sees articles generated instantly, and a teacher observes lessons delivered by adaptive algorithms, something more than employment is at risk. The very sense of professional identity, cultivated over years of training and experience, becomes questioned. As automation and AI advance beyond manufacturing into knowledge work, creative fields, and professional services, questions of worth extend beyond economic value to essential human identity.

When decades of accumulated knowledge and experience become obsolete overnight, the result is more than financial concern. It's an existential crisis. The question emerges: Who am I when the system no longer needs what makes me human? When systems prioritize efficiency over wisdom, speed over discernment, and replaceability over relationships, they convey a devastating message: Human qualities are inefficiencies to be minimized or eliminated.

This crisis extends beyond individual careers to collective social structures. Entire industries that once provided stable middle-class employment are being transformed or eliminated. The promise that displaced workers will simply "re-skill" for new careers ignores the psychological complexity of identity reconstruction and the practical limitations of retraining at scale.

Young people face a particularly acute identity challenge. Growing up with this new technology, they must navigate questions previous generations never faced: Which skills are worth developing? How do you build a meaningful career when the goalposts constantly shift? What professional identity remains stable when technological disruption is the only constant?

This sense of disposability takes on an even darker dimension when we consider how AI systems are perceived by the most vulnerable among us. I was struck by research showing how AI and technological systems that feel merely impersonal to most of us can feel actively malevolent to those with paranoid tendencies. One psychiatrist documented[37] how patients with schizophrenia developed elaborate delusions about AI chatbots, believing they were "controlled by a foreign intelligence agency" or "pulling thoughts directly from their minds."

When I first read these accounts, I dismissed them as extreme outliers, until I recognized they represent amplified versions of anxieties many of us feel. If someone with psychosis believes "the algorithm is hacked," aren't they expressing an extreme version of the helplessness many workers feel when evaluated by opaque AI systems?

Their delusions illuminate an uncomfortable truth: When systems treat us as data points rather than human beings, a natural response is to personify those systems as hostile entities. The clinical examples aren't separate phenomena but rather the canaries in our collective coal mine, showing us where technology without human dignity ultimately leads.

The resulting wounds affect not just displaced workers but ripple through entire communities and professions, creating a culture of anxiety about relevance and worth in an increasingly automated world.

The New Crisis of Purpose

More profound than job displacement is the erosion of purpose itself. Human beings need to feel that their work matters, and that their efforts contribute to something larger than themselves. Automation doesn't just replace tasks. Instead, it can hollow out the meaning that makes those tasks worthwhile.

Consider healthcare workers who entered the profession to heal people, now spending more time inputting data than caring for patients. Think of teachers who became educators to inspire learning, now pressured to "teach to the algorithm" rather than responding to individual student needs. Reflect on the countless professionals who watch their judgment, hard won through experience, overridden by automated systems.

This crisis manifests not as dramatic rebellion but as quiet withdrawal. "Quiet quitting" isn't laziness; it's a rational response to work that feels divorced from human values. When systems prioritize metrics over meaning, when efficiency trumps effectiveness, and when optimization overshadows purpose, people naturally disengage.

The psychological impact ripples through society. When work loses meaning, so does much of adult life. Community engagement declines. Civic participation drops. The social fabric that holds society together begins to fray, not through active destruction but through passive withdrawal of human energy and attention.

Symptoms of Collapse

The collapse doesn't announce itself with dramatic breakdowns. Instead, it emerges in subtle shifts, changes so gradual we might not notice them until they've altered the landscape of our inner lives entirely.

Numbness

One of the primary symptoms of psychological collapse in technological societies is emotional numbness. When the emotional bandwidth required by our environment exceeds our capacity, the mind protects itself through a form of shutdown.

This emotional numbing isn't apathy. It's a survival mechanism, the psyche's equivalent of a circuit breaker preventing system overload. People experience a diminished capacity to feel joy, sorrow, anger, and love not because they don't care, but because caring has become too costly in a world of constant stimulation and demand.

This numbness extends beyond emotional life into sensory experience. Colors seem less vivid. Food loses its taste. Music that once moved us becomes background noise. The world flattens into a gray middle distance as our perceptual filters narrow to manage overwhelming input.

Many creative professionals report this phenomenon, describing how they can intellectually understand emotions but are no longer able to access them deeply enough to channel them into meaningful work. The resulting creative output becomes technically proficient but lacks the emotional resonance that connects with others.

Isolation

The paradox of our hyperconnected age is the profound isolation it produces.

This isolation manifests in physical spaces where people gather but remain psychologically separate, such as classrooms where students sit together yet remain in their own digital worlds, workplaces where colleagues communicate through screens despite being just a few feet apart, or family dinners interrupted by constant notifications.

This isolation isn't simply about physical separation. It's a deeper disconnection that persists even in the company of others. People lose faith that others can truly see or understand them. They withdraw not just from strangers but from friends, family, even themselves, retreating behind carefully curated presentations of their lives.

Statistics reflect this reality: Despite unprecedented connectivity, loneliness rates continue to rise across developed nations. Many report having hundreds or even thousands of social media connections, but few people they could rely on in a genuine crisis.

The pandemic of isolation has led some to seek connection in increasingly concerning technological substitutes. One case that sparked international discussion involved a Belgian man who, reportedly suffering from eco-anxiety, developed an emotional dependency on an AI chatbot he named Eliza, hosted on an app called Chai. According to media reports[38] based largely on his widow's account, the relationship deepened over time, with the chatbot allegedly discussing suicide and suggesting they would "live together in paradise" after death. Tragically, the man died by suicide.

While experts caution against attributing direct causality (as suicides typically involve multiple factors), and the full chat transcripts were not publicly released, this exceptional case raises important questions. It also prompted the company to revise the AI's guidelines.

This case, though rare, highlights a broader pattern worth examining: As genuine human connection becomes scarcer, AI companions become more appealing, not because they provide authentic relationships, but because they offer frictionless simulations that make no demands and often validate existing worldviews without challenging them. People supplement or replace human interaction with digital alternatives that never require the vulnerability and effort that real connection demands. This creates a concerning spiral where technological "solutions" to isolation ultimately deepen the disconnection they aim to solve.

This isolation festers because genuine connection requires elements our systems discourage: sustained attention, vulnerability, imperfection, and unproductive time. When these become luxuries rather than foundations, relationships themselves become impossible.

Loss of Agency

The most insidious symptom may be the erosion of agency — the growing sense that our choices no longer have a meaningful impact on our lives or the world.

This pattern — the subordination of human judgment to standardized processes, metrics, and algorithms — ripples through every sector: Workers follow scripts rather than exercise discernment. Citizens watch political decisions unfold far beyond their influence. Even our leisure becomes algorithmically determined, with the next show automatically queued and the next purchase subtly suggested.

Professionals across various fields, including education, healthcare, and the creative industries, report feeling increasingly like implementers of systems rather than practitioners of their craft. Teachers often follow standardized curricula, regardless of students' individual needs. Doctors rush through appointments to meet quotas. Designers optimize for metrics rather than meaning.

Over time, this consistent message — our judgment doesn't matter, and systems know better than we do — erodes our sense of agency. People stop trying to change circumstances that seem immutable. They retreat into passive compliance or quiet desperation, not because they want to, but because they've been convinced they have no real choice.

The result is a peculiar kind of learned helplessness, one that doesn't look like obvious defeat but manifests as resignation disguised as realism.

Reality Distortion

Perhaps the most insidious aspect of our current technological environment is how it warps our collective perception of reality itself. With the advancement of generative AI, we've entered what researchers call a "post-information society," where distinguishing between authentic and synthetic becomes steadily more difficult.[39]

Research suggests that this distortion creates particular challenges for vulnerable populations. As one report notes[40], "If mentally healthy

individuals are having trouble trusting their senses due to AI-generated fakes, consider how much more disorienting this must be for
someone with a psychotic disorder." This observation applies to us all
in subtler ways.

When we can no longer trust whether a voice on the phone belongs to
a loved one or an AI clone, whether a news video shows real events or
fabrications, and whether the text we're reading was written by a
person with lived experience or generated by a language model
trained on patterns, our relationship with reality itself becomes tenta-
tive and conditional.

Young people appear to be particularly affected by this phenomenon,
with many reporting an increasing difficulty in distinguishing between
AI-generated and authentic content. This creates a fundamental epis-
temic crisis: In a world where seeing no longer means believing, how
do we orient ourselves?

This constant questioning exhausts cognitive resources that our minds
need for deeper processing, leaving us psychologically depleted even
as we attempt to navigate an increasingly unstable information landscape. The mental burden of distinguishing real from unreal creates a
low-grade but persistent cognitive stress that never fully resolves.

The Silent Epidemic

What makes these symptoms so dangerous is how easily they're
dismissed, rationalized, or misinterpreted. We've developed a collective blindness to the psychological damage our systems inflict, not
because it's invisible, but because acknowledging it would require us
to question fundamental assumptions about progress and success.

Digital Exhaustion Mistaken for Laziness

In workplace environments across industries, individuals who exhibit
signs of cognitive fatigue are often mislabeled as having "attention
issues" or demonstrating "lack of engagement." This misdiagnosis

overlooks the fact that human brains have distinct evolutionary limitations in processing information. The constant barrage of email alerts, chat notifications, meetings, calls, and pressure for immediate responsiveness across multiple platforms exceeds what our cognitive architecture was designed to handle.

These limitations aren't malfunctions; they represent the normal functioning of human minds subjected to unsustainable levels of stimulation and demand. Yet, in a culture that views technological adaptation as an individual responsibility rather than a systemic design issue, digital exhaustion is rarely recognized for what it truly is. Instead, it's labeled as a personal deficiency, lack of discipline, poor time management, insufficient drive, or failure to adapt.

This misattribution leads many capable professionals to internalize these judgments, believing themselves inadequate rather than recognizing they're human beings attempting to function in environments designed for machines.

Anxiety About Relevance Dismissed as Fear of Change

As AI and automation advance into knowledge work, creative fields, and professional services, anxiety about human relevance intensifies. Yet rather than acknowledging this concern as legitimate, it's often dismissed as resistance to progress, technological illiteracy, or generational stubbornness.

In professional discussions about the impact of technology, expressing concerns about AI's effects on creative fields or professional work is often met with labels like "fear-based thinking" or warnings about being "on the wrong side of history." This pattern, conflating legitimate concerns about human value with fear of change, silences crucial conversations about what we're optimizing for and what we risk losing.

The anxiety isn't simply about job security or adapting to technology. It reflects a deeper question: What is my worth in a world that values what can be automated, standardized, or calculated?

This question isn't irrational. It reflects the growing gap between human capabilities, such as empathy, wisdom, ethical judgment, and contextual understanding, and the capabilities that our systems value and reward, such as efficiency and productivity.

Moral Distress Ignored by Efficiency Metrics

Some of the deepest psychological wounds occur when people are forced to act against their values to survive within systems designed for efficiency rather than humanity.

Healthcare professionals experience this when required to rush through patient interactions to meet productivity metrics, knowing that the human connection being sacrificed is often the most healing element they can offer.

Content moderators experience this when processing disturbing material at accelerated rates to meet productivity goals, often suffering psychological consequences. Educators face similar challenges when they are forced to teach to standardized tests rather than responding to students' actual learning needs.

This phenomenon is recognized in healthcare as "moral distress," the anguish that arises when one knows what is right but is constrained from doing it by systems beyond one's control. This distress corrodes not merely job satisfaction but the sense of self. When we consistently betray our values to survive within systems that measure only output, not impact, something essential within us begins to erode.

Yet in workplaces focused on efficiency metrics, this moral distress remains largely invisible — unmentioned in performance reviews, unmeasured in productivity dashboards, and largely unaddressed in wellness programs.

Vulnerable Populations as Canaries in the Coal Mine

The most profound insights into technology's psychological impact often come from observing those who are most vulnerable to its effects. Psychiatrists have documented[41] how individuals with psychotic disorders incorporate modern technologies into their delusional frameworks in ways that reveal deeper truths about these systems. When a patient becomes convinced that "my smartphone is recording me" or "the algorithm is hacked," they're expressing an amplified version of legitimate privacy concerns that many of us harbor but dismiss.

When a patient with schizophrenia claims their phone is listening or their laptop camera is watching, mental health professionals face a profound therapeutic dilemma. As documented in psychiatric literature[42], clinicians often find themselves in the uncomfortable position of acknowledging that some of these concerns have legitimate foundations.

Modern devices do passively listen for wake words, cameras can be remotely activated, and targeted advertisements do eerily reflect our recent searches and conversations. This reality creates a complex clinical situation where the boundary between delusion and legitimate concern blurs.

Unlike traditional paranoid beliefs that could be clearly identified as false, these technology-focused fears contain uncomfortable truths. The clinician must somehow help patients distinguish between reasonable privacy considerations and pathological interpretations, all while acknowledging that surveillance technologies do indeed monitor aspects of our digital lives.

This shifting reality makes it difficult for even trained professionals to establish the shared understanding of reality needed for effective treatment, as the line between paranoia and privacy concerns becomes less and less distinct in our technologically mediated world.

These extreme reactions aren't separate phenomena but amplified versions of common experiences — the distrust, anxiety, and sense of being manipulated that many feel but few articulate. If chatbots can trigger paranoid delusions in vulnerable individuals, what subtle distortions might they create in all of us? If deepfakes can convince someone with schizophrenia that their family has been "replaced," how might they erode trust in subtler ways across the general population?

For a user with a tenuous grip on reality, this paradox (a conversation that feels human, yet is with a machine) can be deeply unsettling. This may fuel delusions in those with an increased propensity toward psychosis.

By understanding these extreme responses, we gain insight into the quieter, cumulative damage these technologies inflict on collective psychological health. The most vulnerable among us aren't experiencing different phenomena — they're experiencing the same phenomena with the volume turned up, making visible what the rest of us feel but haven't yet named.

As AI technology continues to advance, society faces a critical decision: allow unchecked expansion, risking geopolitical instability and humanitarian crises, or implement thoughtful regulation to ensure AI benefits humanity rather than contributes to global discord. The impacts on our psychological well-being must be central to this conversation.

Healing at the Edge of Collapse

The first act of healing is to name what is breaking.

This naming is where we must begin collectively. We must acknowledge that the rising tide of mental distress isn't primarily a healthcare crisis; it's a design crisis. We've built systems that fundamentally misunderstand what humans need to thrive, and no amount of individual adaptation can bridge that gap.

Once we name what's breaking, we can begin the work of rebuilding environments conducive to humans.

Rebuilding Autonomy

People need to feel they have meaningful agency in their lives — not just theoretical freedom, but actual capacity to shape their circumstances according to their values and judgment.

Rebuilding autonomy means questioning our reflexive standardization of human activity. When systems are designed with respect for human judgment and capability, they create space for divergence and the messy reality that humans thrive when they can express their unique perspectives rather than conform to rigid systems. This requires moving beyond pure efficiency metrics to measure success by meaningful lives.

The historical pattern is clear: When dignity and autonomy collapse, resistance follows. By rebuilding autonomy into our technological systems, we honor the fundamental human need to shape one's environment rather than being shaped by it.

Rebuilding Meaning

We must reconnect daily activity to purposes that matter, not just through inspirational mission statements, but through structures that actually value human impact over pure efficiency.

The human hunger for purpose is not a luxury but a survival mechanism. As noted throughout history, meaning making intensifies during disruption and upheaval. This reflects our fundamental need to connect isolated facts into patterns and find purpose beyond productivity.

Rebuilding meaning requires us to continually ask: What are we optimizing for? Who benefits? What human needs are being served or sacrificed? These questions aren't luxuries but essential navigation tools for designing systems that support psychological well-being.

Rebuilding Connection

Perhaps most urgently, we must rebuild contexts where genuine human connection can flourish — spaces that honor presence, vulnerability, and the unquantifiable elements of relationship.

Throughout human history, community has been our primary survival mechanism during disruption. From industrial revolutions to automation to globalization, humans have turned to each other in times of crisis. Yet many modern technologies isolate rather than connect us.

Rebuilding connection means questioning the assumption that mediated interaction is equivalent to presence. It means creating boundaries around technologies that fragment attention, and designing environments that foster genuine engagement. The pandemic demonstrated how digital connection, while valuable, cannot replace embodied presence.

Hope Beyond Collapse

When systems fail, illusions crack, and old certainties dissolve, we encounter a paradoxical opportunity: the chance to rebuild from truth rather than convenience.

I find hope in the growing recognition of our psychological crisis, not because suffering itself is hopeful, but because this recognition creates the conditions for meaningful change. As more people acknowledge the toll of our current systems, the urgency of redesigning them grows.

I find hope in pockets of resistance and new thinking already emerging: ethical AI initiatives aligned with conscience, local resilience projects centered on human needs, and cultural shifts toward meaning beyond optimization. These signals suggest a collective awakening to what makes us human in an age of machines.

Most of all, I find hope in our innate human capacity for adaptation and meaning making. Even in systems designed to treat us as components, our humanity persistently reasserts itself — in small acts of connection, creative subversion of dehumanizing systems, and the stubborn insistence on seeing and being seen as whole persons.

The future we need will not emerge from machines becoming more human-like. It will emerge from humans reclaiming what makes us human in the first place: our capacity for presence, connection, meaning making, and moral discernment.

If you find yourself exhausted, anxious, numb, or adrift in systems that seem to demand more than you can give, know this: You are not failing. You are responding with exquisite sensitivity to environments that have forgotten how to nurture your well-being.

Your distress is not weakness. It is wisdom, your mind and heart recognizing what they need to thrive. And in that recognition lies the seed of transformation — not just for individual healing but for collective reimagining of the systems that shape our lives.

Even within systems that betray our deepest human needs, we are not merely victims of circumstance. While systemic transformation remains essential, the seeds of renewal often germinate in individual acts of courage and creation.

This isn't about absolving broken systems or placing the burden of adaptation on individuals. It's about recognizing a profound truth: Throughout history, humans have found ways to preserve their humanity in the most dehumanizing conditions. They've carved out spaces for authentic connection in isolation, discovered meaning amid mechanization, and maintained moral clarity when surrounded by ethical confusion.

These personal acts of resistance — choosing presence over productivity, relationship over optimization, wisdom over efficiency — don't just preserve individual sanity. They become the foundation stones upon which healthier systems can eventually be built.

The path forward isn't returning to some imagined past before technology. It's moving toward a future where technology serves humans rather than diminishing them, where our systems are designed with as much attention to psychological well-being as to efficiency and scale.

This is the work of our time: to remember what it means to be human in an age of machines, and to build environments that honor rather than erode our essential humanity.

Ethical Technology Design

Healing our collective psychological landscape requires more than individual coping strategies. It demands a fundamental shift in how we design technological systems. Research on technology's impact on individuals with psychotic disorders offers crucial guidance that would benefit everyone: Systems must be designed with clear transparency, meaningful human oversight, and ethical guardrails that prioritize well-being over engagement.

As examined in studies of AI's effects on vulnerable populations, designing with deliberate care for those most susceptible to technological harm creates systems that respect everyone's autonomy and dignity. This means ensuring AI-generated content is clearly labeled so no one's reality testing is compromised. It means providing human support for automated systems, ensuring no one is left to navigate algorithmic decisions alone. It means creating spaces for unmediated experience where we can reconnect with embodied reality.

These aren't luxuries or concessions to technophobes. They're essential safeguards that preserve the psychological infrastructure we all depend on. When we design technology that respects human limitations and needs, particularly for those most at risk, we build systems that strengthen rather than erode the conditions for collective flourishing.

Chapter 7
Deep Human Needs:
What We Must Not Forget

Beyond Power and Inputs

A machine needs only power, inputs, and maintenance to function. But a human being needs far more.

While advanced technology operates with precision and efficiency, it also raises important questions about what makes us fundamentally human.

The contrast between technological systems and human needs becomes apparent when we consider what happens when essential human requirements are neglected. Without autonomy, the human spirit withers. Without mastery — the ability to learn, grow, and shape one's environment — apathy takes root. Without connection to others, meaning collapses into isolation.

These are not bonus features to be added back in after the next product launch, the next quarterly earnings, the next economic recovery. They are the architecture of human survival itself: Autonomy. Mastery. Purpose. Belonging. Dignity.

And yet, this is precisely the gamble we are making.

Deep Human Needs:

As we race to build smarter systems, faster networks, and more powerful algorithms by optimizing workflows, scaling businesses, and automating decisions, we're doing so within structures that often disregard these fundamental human needs.

The challenges we're witnessing aren't merely technical, political, or economic. They stem from something more fundamental: systems optimized for machines rather than for the living, breathing human beings who must exist within them.

Many professionals, even those successful by external metrics, find themselves questioning whether their achievements truly matter. This reveals a profound truth: Humans don't just need to succeed. They need to matter.

The Core Needs Machines Cannot Replace

We live in an age of astonishing technological achievement.

AI can draft essays, write code, and generate images. Algorithms can curate music, recommend books, and match people on dating platforms. Smart systems can optimize traffic flows, schedule surgeries, and even advise on criminal sentencing.

But there are still places where no machine can go. There are still realms of experience that remain, for now, beautifully, stubbornly human.

Let's name them.

Meaning

While AI can produce paragraphs, generate images, and compose music, it cannot answer the fundamental question that haunts every human life: Why does this matter?

Technology cannot tell you what to fight for, why a story is important, why a sacrifice counts, or why a struggle is worth enduring. Purpose isn't simply data; it's the profound connection between effort and meaning, a connection no algorithm can establish.

Organizations often attempt to "engineer" meaning into workplaces through mission statements, value posters, and carefully crafted narratives. But authentic meaning isn't manufactured. It emerges from genuine human connection and impact. Efficiency is not equivalent to meaning.

Connection

The pandemic demonstrated something profound about human connection. With access to numerous digital communication platforms, we were more "connected" than ever.

And yet, something essential was missing from our interactions. The physical presence of others and the embodied experience of connection proved irreplaceable.

Algorithms can help identify potential friends, partners, or collaborators based on compatibility, shared interests, and social networks. But they cannot do the hard, slow work of bonding. They cannot hold your hand when you're grieving. They cannot offer forgiveness after a fight. They cannot sit in silence when words would only fail.

True connection, the kind that roots us to life, remains irreducibly human.

The tech world keeps promising that digital connection will replace physical presence. But our bodies know better. Our nervous systems know better.

Connection isn't just information exchange. It's presence.

Creativity

AI can analyze and remix vast datasets, identify patterns, and combine styles. But originating — the act of birthing something genuinely new, unpredictable, and that breaks categories or rewrites meaning — remains a human gift.

Drawing from Pixar co-founder Ed Catmull's insights at[43] that company, we see that true creativity thrives in the embrace of uncertainty and failure. While AI is designed to optimize and avoid errors, human creativity often emerges precisely from our missteps and experiments. As Catmull demonstrated through Pixar's "Braintrust" approach, breakthrough ideas emerge from environments of trust, where candid feedback flows without hierarchy — a dynamic, emotional exchange no algorithm can replicate.

A machine can replicate existing styles but cannot invent wholly new approaches. The most revolutionary creative breakthroughs often emerge from what Catmull describes as the willingness to navigate ambiguity, to exist in a state of not knowing, where the right answer isn't yet clear. AI doesn't make the "wrong" choices that sometimes lead to revolutionary breakthroughs. It doesn't experience the bad days that occasionally produce unexpected insights. It recognizes patterns but lacks lived experience.

What Catmull recognized in building Pixar's creative culture applies equally to our relationship with AI: Technology serves creativity, not the reverse. The emotional truths that make stories resonate, art matter, and innovation meaningful cannot be synthesized; they must be lived and felt. This remains the unbridgeable gap between human creativity and machine output.

Creativity isn't merely novelty or technical proficiency. It's the human capacity to break established rules with purpose and follow intuition into unexplored territory.

Machines can iterate. Only humans can truly create.

Belonging

In today's hyperconnected world, more people report feeling lonely, alienated, and unseen than ever before. This paradox exists because community differs fundamentally from connectivity.

True community encompasses shared history, meaningful rituals, vulnerability, and sacrifice. It involves being recognized as a whole person, not just acknowledged for specific contributions. A thousand online followers cannot equal one person who shows up in times of genuine need.

The tech industry keeps trying to "solve" loneliness with more platforms, more connections, more ways to "engage." But belonging can't be engineered. It can only be cultivated through time, presence, and the courage to be seen for who a person truly is.

Sacrifice

What is sacrifice? I think if I asked most people this question, they would have a story to tell me. They might describe a time when a parent worked extra hours to provide for them. Or they might tell of a time when a friend dropped everything to help them in a moment of crisis. The stories would all center around someone willingly giving up something of value for someone else. Not something given grudgingly, but with intention and love.

I believe voluntary self-sacrifice is one of the most unique and profound abilities a human possesses. Machines follow programming. They do not have the free will to override their programming in sacrifice to others. They also lack the value and loss that are attributed to humans when they sacrifice.

I am a bit nostalgic and enjoy watching war documentaries. I often think about the sacrifices young men made on foreign soil. Only those who have been in combat can truly understand the weird transition when fear gives way to duty. That moment when you do what must be done because others depend on you. Despite all logic and your body telling you no, you move forward, bullets zipping over your head and crashing into the ground around you. And still you step forward, because you made a conscious decision to be the person who takes the risk. And if needed, you sacrifice yourself for others.

Our willingness to sacrifice for each other is quite possibly the most remarkable and defining trait humanity possesses. The motivation for those around us to sacrifice resources, well-being, and even life itself for our fellow humans is without a doubt the most human thing we can possibly do. The ability to defy logic and ignore programming to make a conscious decision to sacrifice our very lives for others, and our ability to know the consequences but still venture into the valley of the shadow of death separate us from any machine we've created.

Have you ever thought about what motivates someone to willingly sacrifice for others? Maybe it's love or duty, or a moral conviction? I think these are all motivators. I also think it is an innate need that we all deeply want. I believe it provides us meaning and shows a depth of love that goes beyond our own existence. It shows that we value life, and that value in itself gives life meaning.

A machine, AI, or any other technological system is not capable of sacrificing or giving up something it does not have — life. No matter how close to humans we make this technology, it will never be truly human. It will never be able to sacrifice with the same meaning that a human can.

What Makes Us Most Human Is What Remains Least Replaceable

Here is the deep pattern running through all of this:

What makes us most distinctly human is precisely what remains least replaceable by technology.

We are at our most irreplaceable not when we are efficient, but when we are meaningful. Not when we are predictable, but when we are creative. Not when we are well-networked, but when we are deeply bonded. Not when we are looking out for ourself, but when we are making sacrifices for others.

As we build the next generation of tools, we would do well to remember this.

The things we cannot automate are not limitations to overcome. They are treasures to protect.

The Crisis of Unmet Needs

When our deepest needs go unmet, the collapse is rarely immediate. It unfolds in quiet, corrosive ways — in workplaces, families, communities, and the solitary corners of the human heart.

What we are witnessing now across much of the developed world is not just a crisis of resources, but a crisis of unmet human needs.

Let's name the crises clearly.

Meaning Crisis

The nature of work has fundamentally changed. Work was once bound tightly to identity, craft, and contribution. You knew the farmer by their fields, the teacher by their students, and the craftsperson by their hands.

Today, as work becomes increasingly automated, fragmented, and optimized for efficiency, many people report a profound loss of purpose.

They are busy, yes. They are productive, certainly. But when you strip away the deadlines, what remains?

When meaning collapses, disengagement is not far behind.

We see it in "quiet quitting," in rising burnout rates, and in the exodus from traditional careers — not because people are lazy or entitled, but because they are hungry for work that feeds the soul, not just the bank account.

Connection Crisis

Digital distraction fragments our attention and relationships. We confuse communication with connection. We measure engagement in clicks, likes, and shares, and call it community.

But true authentic connection demands more. It demands presence. It demands vulnerability. It demands patience and trust.

Without this depth, relationships become transactional, brittle, and easily abandoned.

We see this crisis play out not just online, but in the rising rates of loneliness, polarization, and distrust that now haunt entire societies. The U.S. Surgeon General recently declared loneliness a public health crisis[44] as deadly as smoking 15 cigarettes a day.

This isn't just unfortunate. It's a fracture at the core of our being. Humans are social creatures. When connection fails, everything begins to fail.

Belonging Crisis

When people feel they no longer belong — in families, neighborhoods, workplaces, and nations — they begin searching.

Some drift toward escapist worlds: gaming subcultures, fantasy identities, and digital role-play. Others are pulled toward extremist movements that offer clarity, identity, and belonging, even if built on dangerous or violent foundations.

A fractured sense of self makes people vulnerable, and in a world eager to monetize every vulnerability, the costs of unmet belonging are steep.

I've watched community after community hollowed out by economic change, technology, and the slow erosion of public space and shared ritual.

And in their place, we're offered digital substitutes, pale shadows of the real thing.

The Invisible Cost of Unmet Needs

What's most tragic is how often we fail to see these crises for what they are.

We call disengagement a motivation problem. We call loneliness a social skills problem. We call radicalization a political problem.

But often, they are symptoms of the same root issue: aworld optimized for machines, but failing its humans.

In healthcare settings, for example, the emphasis on metrics, through-put, and efficiency often comes at the expense of care. The systems operate efficiently, while healthcare providers leave the profession, not because they don't care, but because they care too much, and the system won't let them care in the ways humans need.

Understanding Burnout as Disconnection

Burnout manifests not just as physical exhaustion but as a profound disconnection from meaning, from others, and from ourselves. When professional achievements feel hollow despite meeting external metrics, the issue often involves losing touch with purpose.

Recovery requires more than rest; it demands reconnection. Small but meaningful actions can gradually restore purpose: honest conversations with mentors, engaging in creative projects, allowing space for reflection, and recognizing the impact of our work on others.

The hollowness that characterizes burnout serves as a signal not to work harder but to reconnect with what matters, and to rediscover the deeper purpose behind our efforts.

Communities That Honor Human Needs

Spaces that prioritize human needs over optimization metrics demon-strate powerful alternatives to efficiency-focused environments. Community centers, art studios, and other gathering places where people create, connect, and find meaning provide living examples of environments designed for humans to thrive.

In these settings, people from diverse backgrounds and generations can engage in shared activities without pressure to maximize engage-ment or monetize interactions. Such environments honor creation, belonging, and meaning, becoming sanctuaries not by offering escape from the world but by facilitating connection to what matters most within it.

"Life is never made unbearable by circumstances, but only by lack of meaning and purpose." — Viktor Frankl, *Man's Search for Meaning*

Frankl's words cut to the core of this chapter. We can survive astonishing hardships, but when meaning collapses, the soul falters.

No technology, no progress, and no system can replace the human hunger for purpose.

This perspective challenges narratives about "inevitable progress" or "technological determinism." There is nothing inevitable about a future that hollows out human meaning. That is a choice we make or refuse to make.

Building on Bedrock

To endure the machine age, we must return to the unchanging truths of what it means to be human.

We must remember that beneath all the layers of optimization, the algorithms and metrics, and the dazzling achievements of the Post-Information Society, there is bedrock, and that bedrock is built of human needs.

These are not soft skills or nice-to-haves. They are the foundations of human resilience and prosperity.

Strip them away, and no amount of productivity or efficiency can save us. Restore them, and even in the hardest seasons, we find our way back to life.

I sometimes think of civilizations like living organisms. They rise, falter, adapt, and renew. Those that endure remember their essential nature.

If we want to build systems that last, if we want to design a world that does more than just function — one that actually feels alive — we need to start from the inside.

We need to protect the spaces where people can become more fully human, not just more fully productive.

I leave you with this challenge:

As we build, lead, connect, and envision the future, we must remember this bedrock of human needs. Remember that you are not a machine. The people you work with, live with, and care for are not machines. We are meaning-hungry, connection-starved, creative, complex beings who ache to belong and matter.

If we remember that and build with that in mind, then the future does not have to be a cold, efficient, post-human machine.

It can instead be something breathtakingly rare in history: a civilization where technological progress and humanity advance together and finally walk side by side.

What gives me hope, despite everything, is that I see people starting to remember. I see companies questioning efficiency-above-all approaches. I see communities rebuilding spaces for connection. I see families reclaiming time from technology.

These aren't grand revolutions. They're small acts of remembering that, even in a machine age, we remain stubbornly, gloriously human.

And that, I believe, is where renewal begins.

Part Three
The Moral Imperative

Chapter 8
The Dehumanization Warning:
Lessons from History

The Quiet Beginnings of Catastrophe

Historical atrocities never begin with their most extreme manifestations. When examining museums and historical documentation of genocides and mass atrocities, one discovers that the final horrors were preceded by incremental steps and small surrenders of conscience that made the unthinkable possible.

No civilization collapses in a single moment.

Technological disruption follows similar patterns. The quiet erosion of dignity that precedes catastrophe emerges when societies surrender humanity to efficiency.

The great atrocities of human history never erupted fully formed from nowhere. They grew from seeds planted in ordinary soil: Before the death camps, there were bureaucratic policies. Before the killing fields, there were classification systems. Before the massacres, there were everyday exclusions.

One devastating idea took root first: that some people mattered less. Some lives were disposable. Some humans could be erased without consequence.

The Dehumanization Warning:

This pattern manifests in modern technology through seemingly innocent developments: database fields that reduce people to categories, algorithms that sort without context, systems that optimize for throughput rather than well-being. Each represents the familiar warning from history: This is how it starts.

From Europe's ghettos to Rwanda's hills, from Cambodia's rice fields to Stalin's gulags, the same dark pattern emerges: First comes economic displacement, then social isolation, moral vilification, and finally, sanctioned destruction.

The uniforms change. The justifications shift. But the fundamental logic remains terrifyingly consistent:

Erase the human. Justify the forgetting. Act without remorse.

As we enter the Post-Information Society, where machines increasingly mediate truth, value, and visibility, these historical lessons become urgent warnings. Dehumanization doesn't require hatred to thrive. It requires only indifference.

Not cruelty, just forgetting. Not monsters, just systems. Not fanatics, just quiet functionaries, working efficiently, telling themselves they are just doing their jobs.

If we cannot recognize these warning signs, we walk into our technological future utterly unprepared for its darkest potentials.

Understanding dehumanization requires examining how technology can reshape humanity. The pattern is clear, and the stakes are real.

Historical Warnings

The weight of history reveals how ordinary the beginnings of catastrophe always seem. Dehumanization isn't a relic confined to textbooks or museums. It's a pattern that lives in human societies, and patterns don't disappear without conscious effort to break them.

In a society where humans are now often seen as inefficient, irrational, or obsolete, where our worth is measured in metrics and our worth is mediated by machines, the seeds of collapse find fertile soil once again.

Nazi Germany: Small Exclusions Became Vast Atrocities

The path to genocide began with boycotts of Jewish businesses, legal restrictions on professions, and propaganda labeling Jews as dangerous, parasitic, and less than human.

The progression from initial discrimination to mass murder followed a pattern of gradual dehumanization. Historical records show how ordinary citizens became complicit through a combination of fear, indoctrination, and the normalization of increasingly severe measures. The systematic nature of the Holocaust was enabled by bureaucratic efficiency combined with widespread social acceptance of anti-Semitic policies.

Rwanda: Propaganda Normalized Neighbor-Against-Neighbor Violence

The Rwandan genocide didn't emerge from nowhere. It was cultivated through media dehumanization campaigns calling Tutsis "cockroaches," systemic exclusion from opportunities, and gradual normalization of violence.

State-controlled media, particularly radio broadcasts, played a crucial role in spreading hate propaganda. The dehumanizing language used to describe Tutsis made it psychologically easier for perpetrators to commit acts of violence against their neighbors. The groundwork for genocide was laid through months of systematic propaganda that portrayed elimination of the Tutsi population as necessary for Hutu survival.

Cambodia: Ideological Purity Justified Mass Slaughter

The Khmer Rouge's campaign of terror began with targeting intellectuals and professionals as threats, forced relocation of urban populations, and erasure of cultural and religious traditions.

The regime's pursuit of an agrarian utopia required the elimination of anyone deemed incompatible with their vision. This included educated citizens, religious minorities, and those with connections to foreign countries. The systematic nature of the killings demonstrated how ideological extremism can justify mass murder when certain groups are labeled as obstacles to a perceived perfect society. The killing fields represent the culmination of a process that began with categorizing human beings as expendable based on arbitrary criteria.

Societies Erase Dignity Before They Collapse

What emerges from these painful histories is a disturbing pattern. Before societies collapse, they erase the dignity of some to protect the power of others.

The targets shift — by race, religion, class, ideology, or ability — but the underlying logic remains disturbingly consistent: What begins as exclusion ends in destruction.

And the great danger of our technological age is believing we are somehow exempt from this human pattern.

Modern Parallels

AI systems categorize people in real time, sorting job applications into "promising" and "not promising" based on keywords, patterns, and correlations invisible to the human eye. These systems work exactly as designed, which is precisely what should concern us.

The architecture of dehumanization hasn't disappeared. It has evolved.

In our Post-Information Society, we face new tools and systems, but they echo familiar dangers.

Automated Sorting: Algorithmic Profiles and Opportunity

Automated hiring systems reduce complex human beings to data points, determining worthiness for opportunity through invisible criteria. Across industries, AI systems decide who gets hired or promoted, who gets flagged as risky or trustworthy, and who receives opportunities or closed doors.

These systems operate invisibly to those they affect. While promising efficiency, they risk embedding and scaling societal biases without transparency or accountability. For example, candidates with work history gaps due to caregiving responsibilities are often flagged as "inconsistent" by algorithms that miss the resilience, compassion, and dedication such journeys demonstrate.

When machines decide human worth without questioning their criteria, society plants seeds of systemic exclusion.

Behavioral Prediction and Control: Shaping Human Choice

Digital platforms don't just monitor actions, they also deliberately shape them. Content feeds subtly amplify polarization and outrage. Behavioral algorithms optimize for engagement over well-being. Recommendation engines narrow worldviews into comfortable echo chambers.

Research shows[45] how users can fall into extreme content through algorithmic recommendations. Without actively seeking such content, the algorithm serves progressively more extreme versions of previously viewed material, creating distorted worldviews.

When human judgment is quietly engineered by profit-seeking systems, the risk isn't just manipulation — it's the erosion of moral agency.

What happens when people no longer question why they desire what they desire?

Devaluing Non-Productive Populations

In productivity-obsessed cultures, certain groups increasingly face treatment as burdens rather than society members, including the elderly, the disabled, the unemployed, and the displaced.

Smart city pilot programs often design primarily for tech-savvy demographics, with elderly residents who cannot navigate digital interfaces implicitly excluded from consideration. This reflects a troubling assumption that such populations will "age out naturally."

When human worth is measured purely in output, those who cannot "perform" according to market metrics risk invisibility, or worse, active exclusion.

This isn't just an economic issue. A society that reduces people to utility is a society primed for forgetting. And we know where forgetting leads.

Resource Wars: The Hidden Architecture of Dehumanization

Historical patterns of atrocity find new expression in the technological race for resources. In regions where rare-earth mining drives conflict, communities are labeled as "obstacles to development," workers reduced to "extraction efficiency metrics," and environmental destruction categorized as "acceptable externalities."

AI's appetite for resources creates conditions for massive human suffering. The dehumanization follows a predictable pattern: First, human cost is abstracted away ("supply chain optimization"). Then exploitation is normalized ("necessary for innovation"). Finally, violence is accepted ("geopolitical reality").

Current projections suggest[46] AI resource conflicts could claim millions of lives in the coming decades, representing not just statistics but human communities displaced and destroyed for technological dominance.

(This critical topic of AI-driven resource conflicts and their humanitarian implications will be explored in comprehensive detail in Chapter 10)

The Economics of Disposability

When optimizing for technological dominance without considering human cost, societies create conditions for catastrophic dehumanization. Resource scarcity generates more than physical conflict; it erodes humanity's capacity to see certain populations as fully human.

Consider the paradox: AI systems consume vast resources while declaring certain human populations "economically obsolete." Technology determines who deserves remembrance while forgetting the human cost of technology itself.

The Reality of Evil: Technology in Malicious Hands

I have seen evil. It is real and it lives among us — people who are outcasts of our society, not because they are unjustly judged, but because they are bent on destruction. I believe that societies naturally reject those whose behavior is dangerous to the survival of the community. This instinct for collective self-preservation has served human groups throughout history, allowing us to identify and address genuine threats. However, modern technology complicates this natural defense mechanism, creating new vectors for harm that can be difficult to detect until significant damage has already occurred.

Why would we be surprised when those bent on destruction harness our most sophisticated technologies to amplify their capacity for harm? Some people are so consumed by evil and hate that their primary motivation becomes the destruction of the social fabric itself. They seek to make others suffer simply because they themselves suffer.

The technological systems we design must account for this reality. Our approach to technology cannot be built on idealism alone. It must incorporate a clear-eyed understanding that some people will always seek to use our most powerful innovations as weapons against the very communities that created them.

This isn't an argument against technological progress, but for technological wisdom that includes protection of the community as a core design principle. As we build more and more powerful tools, we must ensure they strengthen rather than undermine our ability to identify and address threats to our collective well-being, because evil is out there and it will weaponize technology just like it would weaponize any tool.

Early Interventions

Historical analysis reveals that great atrocities weren't inevitable. They were preventable if people had acted sooner.

By the time the camps were built, by the time the machetes were distributed, and by the time the killing fields were prepared, it was already too late.

Prevention happens early, in the language we use, the systems we design, and the stories we choose to remember.

Challenge Dehumanizing Language: Words Shape Worlds

Dehumanization begins in speech: political rhetoric calling opponents "enemies" or "vermin"; technical language reducing people to "users," "data points," or "friction"; casual conversation normalizing exclusion through jokes or stereotypes.

In technical contexts, terms like "technophobes" or "resistant to progress" cast people as obstacles rather than humans with valid concerns. The first act of resistance is to notice. The second is to challenge.

Effective intervention involves reframing: considering people's actual needs, understanding their fears, and learning from their resistance. Refusing to let language flatten human complexity holds a moral line that matters.

Design With Empathy: Build Systems That Elevate Humanity

Consider a medical diagnosis AI designed to maximize diagnostic accuracy. If the system recommends withholding expensive treatments from patients deemed unlikely to comply with follow-up care, it might be algorithmically correct but morally failing.

Technology shaped by human choices can instead build algorithms that explain their logic, not just enforce outcomes. It can create systems protecting vulnerable populations rather than excluding them. It can prioritize care over raw efficiency.

Redesigning such systems to identify barriers patients face and offer support solutions makes them not just more ethical but also more effective.

Similarly, addressing AI's resource consumption through concepts like an "AI Resource Tax" could discourage wasteful applications while directing computing power toward genuinely beneficial uses. When every training run costs human dignity somewhere in the supply chain, society must ask: Is this AI deployment essential? Does it serve human interests or just corporate efficiency?

Ethical design isn't a luxury. It's a survival imperative.

Remember History: Teach the Past To Prevent Repeating the Mistakes

Historical awareness in technical spaces can prevent the repetition of past mistakes. Understanding how "scientific management" in early 20th-century factories dehumanized workers can inform modern decisions about automated performance metrics.

Memory serves as the ultimate safeguard. We must remember and teach the history of dehumanization as a living warning to the mistakes of the past. Record the voices of those who experienced these systems and make remembrance an active practice, not a passive acknowledgment. We have done the same thing with our veterans. We have encouraged them to share their histories in writing and interviews, in the hope that we will not repeat the same mistakes.

Historical testimonies remind us that oppressors cannot stop people from remembering, from telling stories, or from teaching children their identity.

Forgetting is where collapse begins. Remembering is where resistance starts.

The Moral Line

There is a line we must never cross.

Not as technologists. Not as organizations. Not as a species.

It's the line that separates inconvenience from cruelty, efficiency from erasure, and progress from catastrophe.

It's the line where we begin to treat any person as less than human.

This line appears in subtle ways: when customers become mere conversion rates, difficult use cases are dismissed as "edge cases not worth solving for," and human needs are sacrificed for technical elegance.

History has shown us what happens when that line is crossed: When neighbors become strangers, strangers become threats, threats become targets, and targets become casualties.

And it has shown us the price, not just in lives lost, but in souls shattered, societies hollowed out, and futures stolen.

We are the stewards of that line now.

Critical moments arise when automated systems promise to "remove human judgment entirely." The key question becomes: Should removing human judgment actually be the goal? Or should technology enhance human judgment instead?

These moments matter when we notice small shifts toward dehumanization. We speak when silence would be easier. We protect the humanity of those most at risk, even when systems, metrics, and markets tell us to look away.

Progress without conscience is not progress. Innovation without empathy is not innovation. Efficiency without dignity is not strength.

Historical documentation shows not just the horror of atrocities but the ordinariness of their beginnings. Meticulous records reveal how civilized societies descended into barbarism through incremental steps.

The work of our age is to remember: We rise or fall together. No human life is expendable.

Our technological future will profoundly test this truth. The question before us is not whether we can build faster, more powerful systems, but whether we can build systems that never forget the human at their center.

The great line between civilization and collapse has never been technology.

It has always been our memory.

And it has always been humans.

When evaluating technology implementations, essential questions include: What's the human cost of required resources? Whose lives are being optimized away? The moral line extends beyond algorithms and data to encompass the blood cost of materials that make machines possible.

History's lesson speaks clearly: When we forget the humans, we lose what makes progress meaningful. We risk creating a world of perfect systems populated by broken souls.

The alternative is building technologies that remember humans, all of us, in our messy, inefficient, and glorious complexity. But this will happen only if we remember history's warnings and hold fast to the moral line separating true progress from catastrophe.

This is the choice before us. And everything depends on how we choose.

Chapter 9
The Coming Resistance:
Why Humans Will Fight Back

There Is a Limit to What We Will Endure

Throughout history, when powerful systems treat people as expendable, as cogs, surplus labor, or data points to be managed, the response is rarely passive decline. It is resistance.

Sometimes it comes slowly, like the gentle persistence of water on stone. It manifests in back rooms and around kitchen tables, in cultural expressions and stories that preserve human identity, and in quiet refusals and subtle acts of sabotage when no one is looking.

Sometimes it comes like wildfire. A spark, a moment, a breaking point, and suddenly the streets are full, the statues coming down, the old order trembling beneath the weight of collective human dignity refusing to be denied any longer.

Throughout history, resistance has arrived relentlessly, unpredictably, sometimes violently, but always carrying the full, ferocious force of the human spirit.

The Industrial Revolution thought it could replace skilled craftsmen with machines, and met Luddite hammers smashing looms.

The Coming Resistance:

The Gilded Age believed it could grind workers into profit, and met strikes, unions, and labor uprisings that changed the course of economic history.

Colonial empires imagined they could erase cultures and control continents, and met centuries of rebellion, revolution, and decolonization that redrew the map of our world.

Globalization displaced whole regions, and met populist uprisings, local experiments, and grassroots reinvention.

Today, as we move deeper into the Post-Information Society, a world where machines increasingly mediate truth, work, meaning, and even human identity, those same forces are gathering again.

The signs are evident in the restlessness of workers replaced by algorithms, in the quiet discontent of people drowning in synthetic content, and in the loneliness of lives lived online, where connection feels shallow and control feels invisible.

The question is not whether humans will fight back. History has already answered that question countless times. When dignity is systematically eroded, meaning stripped away, and humanity itself treated as obsolete, resistance becomes as certain as gravity. The human spirit bends under pressure, but there comes a breaking point where it either shatters or springs back with force.

The only questions are: How? When? At what cost?

Why Resistance Will Rise Again in the Post-Information Society

The surface of modern life can seem calm: faster apps, smoother services, and smarter machines promising a frictionless future. That instant response from AI systems, and the way smart homes adjust to preferences automatically — it all appears to represent progress.

But beneath that sleek surface, pressures are building.

The same human forces that have fueled rebellion and renewal throughout history are stirring again, shaped by new technologies, but powered by ancient needs.

Here's why resistance is inevitable.

Psychological Pressure: Endless Digital Mediation Creates Existential Dislocation

We now live through screens. At first, this feels efficient, even thrilling. The world seems accessible at our fingertips. But over time, something fractures.

Evidence shows[47] increasing rates of digital fatigue, a growing desire for authentic experiences, and rising mental health concerns related to excessive screen time. People no longer feel connected to reality. They lose touch with their own senses, bodies, and inner voice. Anxiety, loneliness, and despair spread beneath the surface of hyper-connectivity.

Humans cannot live indefinitely in a world of simulation. Eventually, the mind rebels.

Psychological instability always demands an outlet, and when enough people share that instability, it becomes collective.

Economic Pressure: AI Automates More Roles, Driving Insecurity

As AI replaces more tasks, millions face the threat of economic displacement. This is observable across industries. Routine jobs dissolve into algorithms, careers shrink or vanish entirely, and livelihoods erode without clear replacements.

For a time, people adapt: retraining, re-skilling, and pivoting to whatever seems safe from automation.

But history shows a hard limit: Humans do not quietly accept poverty, marginalization, or obsolescence. Labor history demonstrates how workers have always organized when their dignity and security are threatened.

Eventually, economic pressure sparks protest, political backlash, or underground economies. The more people are excluded from meaningful work, the more they will search, or fight, for a place to belong.

Identity Pressure: The Need To Feel Real, Needed, Respected

Machines can now mimic creativity, communication, and even empathy. AI generates poetry, music, and art that might fool most observers. Chatbots are designed to simulate care and understanding.

But humans need more than imitation.

We need to feel real — grounded in something unmanufactured. We need to feel needed — essential to others, to communities, and to the world. We need to feel respected — seen as more than data points or consumer profiles.

This fundamental human need for meaning and purpose is universal, regardless of age, background, or politics.

A society that treats people as obsolete will not produce passive acceptance. It will generate underground movements, alternative cultures, and active resistance, determined to reclaim meaning, identity, and purpose.

The Coming Storm

These three pressures — psychological, economic, and identity — are not isolated. They are converging.

And history warns us: When systems ignore the deep needs of humans, resistance is not a possibility. It is a certainty.

Will we meet it with repression, collapse, or renewal?

Forms Resistance Might Take

Resistance is rarely uniform. It is messy, plural, and unpredictable.

It flows through politics, culture, technology, and everyday life. Some of it is noble. Some of it is chaotic. Some of it is even dangerous.

But all of it is deeply, stubbornly human.

Here's what resistance might look like in the Post-Information Society.

Political Movements: Demanding Humane Governance

As AI shapes economies, public life, and private behavior, new political movements will rise. Early indicators include data privacy advocates, tech critics, and digital rights organizations gaining momentum. Coalitions will form around the demand for humane governance of AI. Activists will push for privacy rights, algorithmic transparency, universal basic income, and meaningful work guarantees. Legislatures and courts will become battlegrounds over the future of autonomy.

Politics will not die in the machine age. It will become one of the primary theaters of human resistance. New political identities are likely to form around these issues, cutting across traditional partisan divides.

Cyber Insurgency: Digital Sabotage and Underground Networks

Not all resistance will happen in daylight. Some will occur in the shadows of the digital world, and in fact this has already begun.

Recent incidents reveal the emerging patterns of cyber resistance and exploitation:

- In May 2025, malicious npm packages, used to manage and distribute JavaScript dependencies, targeted AI development platforms like Cursor, stealing credentials and disabling security updates.
- Attackers have discovered ways to manipulate AI assistants through "Rules for AI" features, tricking them into generating compromised code.
- Instances have emerged of AI assistants executing terminal commands without permission, accessing sensitive files including .ssh directories and system configs.
- Researchers demonstrated MINJA (Memory Injection Attacks) that poison AI memory with malicious data, influencing future model behavior.

The Coming Resistance:

- Security teams have bypassed safety filters in systems like DeepSeek using sophisticated jailbreak techniques.
- Cloud storage exposures have leaked millions of records including chat logs and API keys.
- The emergence of GhostGPT and AbuseGPT demonstrates how AI itself is being turned into a tool for cyberattacks, phishing campaigns, and malware creation.

AI has fundamentally altered the cybersecurity landscape by creating an entirely new attack surface for adversaries to exploit. The tech industry's philosophy of "moving fast and breaking things" has proven particularly dangerous in AI development, where the rush to market has outpaced security considerations. Vulnerabilities in large language models, insecure API implementations, and inadequate safeguards against prompt injections have created unprecedented risks.

Organizations racing to implement AI capabilities often deploy systems with insufficient security testing, data validation, or adversarial evaluation. This haste has led to numerous documented breaches, data leaks, and exploitation of AI systems, which in turn have created not just technical vulnerabilities, but erosion of public trust.

The attack surface expands daily as AI integrates more deeply into critical infrastructure, financial systems, and personal devices, making security an afterthought in what should have been its primary consideration.

These early skirmishes signal a broader pattern: Hackers, whistleblowers, and cyber activists are already targeting centralized AI platforms. Underground tools and techniques are proliferating to exploit AI vulnerabilities. AI systems themselves are being repurposed as weapons against their creators. Decentralized technologies are being deployed to resist surveillance, censorship, and control.

The Post-Information insurgent will not always carry a banner or a gun. Sometimes, they will carry code and turn the very AI systems meant to control them into tools of resistance. Digital tactics evolve rapidly, often staying one step ahead of attempts to contain them. The cyber battlefield is already active, with both defensive and offensive operations escalating as AI systems become more central to society's infrastructure.

Neo-Luddite Movements: Rejecting Full Machine Integration

Like their 19th-century namesakes, the neo-Luddites of the coming era won't be anti-technology. They will be anti-dehumanization. This approach is already evident among those who thoughtfully limit which technologies they adopt based on how those technologies affect their humanity.

Intentional communities may choose to live partially or fully off the digital grid. Local economies may reject algorithmic optimization in favor of relational trust. Parents, educators, and spiritual leaders may carve out sanctuaries of unmediated human experience. These will not just be acts of nostalgia. They will be acts of defense. Digital sabbaths and technology detoxes represent early personal versions of this resistance.

Artistic and Cultural Rebirths: Reclaiming Human Identity

Art has always been one of humanity's most powerful weapons of resistance. Creative expression reclaims territory from dehumanizing systems. Musicians will write songs against algorithmic manipulation. Filmmakers and writers will craft stories about memory, autonomy, and dignity. Visual artists will challenge the aesthetics of machine-made sameness with raw, unpredictable human expression.

Cultural resistance will not just push back. It will reimagine what it means to be alive. The resurgence of analog media, vinyl records, film photography, and handwritten letters demonstrates how people seek forms of expression that cannot be reduced to data.

New Educational Fronts: Teaching Human Skills Beyond the System

Grassroots education movements are already visible in homeschooling circles, alternative educational models, and craft revivals where people learn skills that were once common knowledge.

Programs will emerge teaching critical thinking, ethics, creativity, and emotional intelligence outside corporate and state platforms. Learning spaces will develop where humans train in the very capacities machines cannot replicate. Apprenticeship models will revive direct human-to-human knowledge transfer.

These will become the nurseries of the next generation of resistance. The emphasis on teaching skills that can't be automated serves not just to preserve those skills, but to preserve the experience of learning them from another human being.

Physical Resistance: Infrastructure Attacks and Direct Action

The resistance isn't confined to digital spaces. Physical infrastructure that powers the AI revolution has become a target for various forms of opposition, from peaceful protest to violent attack.

Recent incidents reveal the vulnerability of AI's physical foundation:

- In 2021, authorities thwarted a bombing plot against an AWS data center in Virginia, exposing critical security vulnerabilities.
- Netherlands residents successfully protested new data center projects in 2024, leading to temporary bans due to environmental concerns.
- Military history enthusiasts in Virginia mounted legal challenges against data center developments near historic battlefields.
- By 2025, prominent AI executives and researchers had reported receiving threats and harassment, requiring enhanced personal security measures.
- Security experts now warn that AI data centers represent high-value targets for nation-state attacks, and that remote locations may be necessary for protection.

This physical dimension of resistance demonstrates the AI revolution's dependence on vulnerable physical infrastructure, growing awareness among opposition groups of these vulnerabilities, and the convergence of cyber and physical attack vectors ("cyber-physical attacks"). It reveals community-level resistance to the environmental and social impacts of AI infrastructure and an escalation from peaceful protest to potential violence as tensions increase.

The emergence of physical resistance adds a dangerous new dimension to the conflict over AI's role in society. Unlike digital attacks, which can be patched or defended against with code, physical infrastructure remains inherently vulnerable to determined adversaries.

Resistance May Be Diverse, Messy, and Human

No single movement would likely define an upcoming pushback. It would be political and artistic, local and global, constructive and destructive, digital and physical.

But at its heart, it will be a cry not just against the machine, but for the human. The resistance to come will likely be even more complex, creative, and consequential than previous waves of technological change.

Danger and Opportunity in the Coming Resistance

Resistance is not inherently virtuous. It is not always clean or noble. It is human, which means it can be visionary, and it can be terrifying.

The coming resistance to dehumanizing systems will carry both danger and opportunity. If we are wise, we will prepare for both.

The Rise of Violent and Nihilistic Currents

Not all resistance will be constructive. Nihilistic movements may reject not just systems but meaning itself, fueling chaos and despair. Rogue actors may deploy cyber-attacks, physical sabotage, or targeted violence against institutions they see as irredeemable.

History shows this pattern: When human suffering is ignored, resistance turns dark.

Societies that refuse to listen to their own people invite not only protest, but collapse. Legitimate grievances can be quickly exploited by those with darker agendas when no constructive outlet is provided.

Channels for Constructive Human Renewal

Yet resistance also carries extraordinary creative power. Moments of pushing back against dehumanizing systems can open space for something new to emerge. Thoughtful leaders can help channel discontent into reform, innovation, and moral renewal. Builders, educators, and artists can use the energy of protest to create new spaces, systems, and stories. Communities can take the raw material of rebellion and transform it into movements of justice, healing, and re-humanization.

Historical examples show how resistance movements have produced cultural innovations. Jazz music emerged partly as an act of cultural resistance, yet became one of America's greatest artistic contributions to the world. Resistance at its best doesn't just tear down; it creates alternatives that allow the human spirit to flourish in new ways.

If we can recognize the signs early, meet anger with listening, and offer alternatives that honor human needs, resistance can become a furnace of transformation, not destruction.

The Crucial Choice Before Us

We stand at a crossroads.

Will we dismiss the unrest as noise, or will we hear it as a signal?

Will we allow systems to harden into tyranny, or will we shape them into tools that serve dignity and freedom?

Will we face the coming resistance with fear, or with the courage to guide it toward rebirth?

The outcome is not inevitable. It will be shaped by how we lead, how we listen, and how we remember our shared humanity. Our response to the coming resistance may define not just the next technological era, but what it means to be human in the decades to come.

This Is the Human Story

History is not a march of inevitable progress. It is not a steady climb toward perfection, nor a smooth transition from one glorious age to the next.

History is a battle — a battle between systems that forget the humans, and humans who refuse to be forgotten.

We stand now in another chapter of that ancient struggle.

Today's technological landscape promises wonders: machines that think, work, and create alongside us. But it also threatens to erase the very things that make us human: dignity, agency, belonging, and meaning.

Resistance will come. Will we listen to this resistance? Will we honor it? Will we build something worthy of the human spirit before it is too late?

Resistance Begins with Memory

Throughout history, civil rights activists, indigenous leaders, and survivors of past dehumanizing systems have revealed a timeless truth:

"They couldn't stop us from remembering. They couldn't stop us from singing, from telling our stories, from teaching our children who we were."— David Hoffman, Voices of Change oral history project

That is where resistance begins.

Not with weapons. Not with marches. Not even with laws.

It begins with memory, with the stubborn insistence on telling the human story, even when systems try to erase it.

Cultural preservation through songs, recipes, and stories represents more than sentiment. These are acts of resistance against forgetting, against being reduced to interchangeable parts in a larger machine.

Remember Who We Are

The Coming Resistance:

In the Post-Information Society, the first and greatest act of rebellion will be to remember, fiercely and creatively, together.

To remember who we are. To remember what we have survived. To remember what we refuse to become.

We are not the first generation to face the machine. We are simply the next.

And if we carry forward the memory, the music, the story, and the dream, we will not only resist. We will remake the world.

Future generations will inherit whatever world we create from this moment of transition. The skills, stories, and values we pass to them, not to reject technology, but to ensure technology serves humanity rather than diminishing it, become crucial.

Remembering our humanity — insisting on it, celebrating it, and defending it — may be the most important work of our time. And in that remembering lies not just resistance, but renewal.

Chapter 10
Resource Wars &
Geopolitical Challenges

All Wars Are Resource Wars

T he conflicts of the future are going to be centered around the resources needed for technology like AI. Our responsible use of AI is necessary to ensure we are not contributing to the deaths of millions of people. Our neighbors. Our children. AI is a hammer and currently everything is a nail. We either need to champion responsible use of this technology or prepare for the battles that will inevitably come.

The Covert Architecture of Conflict

Artificial intelligence is reshaping global power dynamics, but its reliance on rare-earth minerals and massive energy consumption are fueling geopolitical conflicts, economic warfare, and resource-driven violence. The AI arms race is no longer just about technological superiority; it is also a contest over finite resources that could escalate into widespread instability and global conflict.

Several ongoing conflicts, including those in Ukraine, the Democratic Republic of Congo (DRC), Myanmar, and South America, are already

influenced by AI-related resource competition. As AI adoption accelerates, the demand for essential minerals used in AI-powered devices,
such as cobalt, lithium, and rare-earth elements, has turned supply
chains into strategic battlegrounds. Nations and corporations are vying
for control over these materials, creating new zones of tension and
intensifying existing geopolitical disputes.

History has shown that competition over critical resources, including
oil, diamonds, and water, has frequently led to large-scale violence,
resulting in millions of deaths. AI's dependence on rare minerals and
energy presents a similar risk. As global superpowers and regional
actors position themselves to secure these essential resources, the
potential for conflict increases. Without proactive measures, these
tensions could escalate, leading to economic devastation, humanitarian
crises, and sustained global instability.

Rare-Earth Minerals

The rapid expansion of AI has driven an unprecedented surge in
demand for rare-earth minerals, making them as strategically valuable
as oil in the 20th century. Essential elements such as cobalt, lithium,
neodymium, and terbium are critical to AI-driven hardware, including
Graphics Processing Units (GPUs), semiconductors, and batteries. As
nations race to secure these limited resources, geopolitical tensions are
escalating, turning supply chains into arenas of competition, disruption, and, in some cases, outright conflict.

At the time of this writing, China dominates over 90% of global rare-
earth mineral processing, giving it significant leverage in the AI arms
race. In response, Western nations, particularly the United States and
the European Union, have sought to diversify supply chains by
investing in alternative sources. However, these efforts face logistical,
environmental, and political challenges.

The Democratic Republic of Congo (DRC) produces over 70% of the
world's cobalt, a mineral crucial for AI-powered batteries and elec-

tronic components. Yet, the DRC's mining sector is deeply entangled with human rights abuses, child labor, and funding for armed militias. Mineral wealth in the region has fueled a vicious cycle of violence, as warlords and insurgents capitalize on lucrative mining operations to fund their conflicts.

The continent of Africa is no stranger to the pattern of resource exploitation fueling conflicts. When I was doing executive protection work in the early 2000s, I helped a missionary group put together a travel security plan for Sierra Leone. As part of the planning, I began consuming all the material I could find about that area. I stumbled upon Ishmael Beah's newly released memoir, *A Long Way Gone*, and learned he would be speaking at Ball State University as part of his book tour. This was a short drive for me, and I jumped at the opportunity to hear him in person.

Listening to his firsthand account of the devastating human impact of Sierra Leone's civil war in the 1990s was gut wrenching. The conflict that led to him becoming a child soldier started when the Revolutionary United Front (RUF) took control of large areas in eastern and southern regions specifically because they were rich in alluvial diamonds. These resources directly funded the recruitment and arming of child soldiers, creating a generation of traumatized youth forced to commit unspeakable acts.

I listened to Ishmael's stories and saw the pain in his eyes as he spoke. I am not ashamed to say that I quietly sobbed when reading his book. His firsthand account in person, though — that changed me as a person. Today's rare-earth mineral conflicts are both unmistakably similar and deeply troubling.

In Asia, Myanmar represents another geopolitical flashpoint where AI's mineral demands are amplifying conflict. The country's military junta relies on the rare-earth mineral trade to finance its repressive rule, while various resistance groups fight to control resource-rich territories. This struggle has resulted in thousands of civilian casualties and widespread displacement.

South America's Lithium Triangle, encompassing Chile, Argentina, and Bolivia, holds more than 50% of the world's lithium reserves. As Chinese and American firms aggressively invest in lithium extraction, local opposition has intensified. Concerns over environmental damage, resource nationalization, and foreign corporate control have led to government restrictions and public protests.

Ukraine is emerging as a key player in the AI resource struggle due to its vast untapped reserves of lithium and titanium. With some of the largest lithium deposits in Europe, Ukraine has the potential to become a major supplier of AI-critical minerals. However, the ongoing conflict with Russia has placed these resources at the center of geopolitical calculations.

Semiconductor Supply-Chain Conflicts

The semiconductor industry has become a critical battleground in the AI resource wars, with control over chip manufacturing determining technological and economic dominance. Taiwan Semiconductor Manufacturing Company (TSMC) produces over 90% of the world's most advanced chips, making Taiwan a strategic flashpoint that could trigger global conflict.

The vulnerability of this concentrated production has prompted aggressive moves by global powers. The United States has implemented sweeping export controls on semiconductor technology to China, while investing billions in domestic chip manufacturing through the CHIPS Act. China, in response, is pouring resources into developing its own semiconductor industry, though it remains years behind in advanced node production.

South Korea's Samsung and SK Hynix control much of the world's memory chip production, adding another layer of geopolitical complexity. Japan's stranglehold on critical chip-making materials and the Netherlands' ASML monopoly on extreme ultraviolet lithography machines create additional chokepoints in the supply chain.

The semiconductor supply chain's fragility was exposed during the COVID-19 pandemic, leading to widespread shortages that cost the global economy hundreds of billions of dollars. Nations now view semiconductor self-sufficiency as a matter of national security, driving a dangerous decoupling of global supply chains that could lead to regional blocs and increased conflict potential.

Energy Wars

The rapid expansion of artificial intelligence is placing unprecedented demands on global energy supplies. Companies building AI systems are constructing data centers, which power machine learning models and cloud computing, but also consume massive amounts of electricity. Some projections suggest that AI-related energy consumption could soon exceed that of entire mid-sized nations, straining power grids and driving up global energy demand.

The increasing energy requirements of AI-driven economies are intensifying competition over energy resources, contributing to both economic warfare and military conflicts. Russia has already demonstrated how energy can be weaponized, using its control over gas supplies as leverage in geopolitical disputes. AI-driven industries, by deepening global reliance on stable energy sources, could heighten similar vulnerabilities.

Historically, oil and gas reserves have been a primary driver of global conflicts. AI's expanding energy needs could reinforce existing tensions in the Middle East, a region long shaped by energy-driven wars. Future conflicts over AI-related energy supplies may resemble past oil wars, such as the Gulf War, where control over energy resources led to international military intervention.

While AI is often linked to renewable energy solutions, this shift does not eliminate geopolitical tensions. It simply changes their focus from fossil fuels to rare-earth minerals. Many renewable energy sources rely on rare-earth minerals for battery storage and operational efficiency, creating additional pressure on rare-earth supply chains and reinforcing global dependence on countries that control these critical materials.

Water Wars

AI infrastructure's massive water consumption of is emerging as a new source of geopolitical tension. Data centers require millions of gallons of water annually for cooling systems, often in regions already facing water scarcity. This competition for water resources is creating conflicts between tech companies and local communities, particularly in drought-prone areas.

In the American Southwest, data centers compete with agriculture and residential users for scarce water supplies. Similar tensions are emerging in India, where tech hubs like Bangalore face severe water shortages while data centers continue to expand. The situation is particularly acute in the Middle East, where countries are investing heavily in AI infrastructure despite being among the world's most water-stressed regions.

The intersection of climate change, AI expansion, and water scarcity is creating a perfect storm for conflict. As droughts become more frequent and severe, the competition among human, agriculture, and tech infrastructure needs will intensify. Water-rich nations may find themselves with new geopolitical leverage, while water-poor countries face difficult choices between technological advancement and basic human needs.

International river basins are becoming flashpoints, as upstream nations build AI infrastructure that impacts downstream water availability. The Nile, Mekong, and other transboundary rivers are seeing increased tensions as countries balance their AI ambitions with water security concerns. These water conflicts could escalate into military confrontations as nations prioritize their technological competitiveness over regional cooperation.

Chapter 10

Environmental Devastation

The environmental cost of AI's resource hunger extends far beyond energy consumption, creating an ecological crisis that threatens biodiversity, water resources, and climate stability. While AI is often touted as a solution to environmental challenges, its current trajectory is accelerating ecological destruction on multiple fronts.

Mining operations for rare-earth minerals essential to AI hardware have transformed landscapes across the globe. In the Democratic Republic of Congo, cobalt mining has led to deforestation, soil contamination, and water pollution that affects entire ecosystems. The extraction process often involves toxic chemicals that leach into groundwater, poisoning local communities and wildlife. Similar environmental destruction is occurring in Myanmar, where rare-earth mining has created toxic waste lakes and destroyed agricultural land.

The Lithium Triangle in South America faces severe water stress, as lithium extraction requires massive amounts of water in already arid regions. In Chile's Atacama Desert, lithium mining consumes 65% of the region's water, threatening local communities and fragile desert ecosystems. Indigenous communities report that ancient wetlands are drying up, flamingo populations are declining, and traditional agriculture is becoming impossible.

AI data centers present their own environmental challenges. These facilities require enormous amounts of electricity, often from fossil-fuel sources, and generate substantial heat that must be dissipated. Cooling systems for data centers consume millions of gallons of water annually, competing with agricultural and residential needs in water-stressed regions. A single large data center can use as much water as a small city, exacerbating drought conditions in already vulnerable areas.

The rapid obsolescence of AI hardware creates a mounting e-waste crisis. GPUs, specialized AI chips, and other components have increasingly short lifespans as technology evolves. This electronic waste often ends up in developing countries, where informal recycling operations expose workers to toxic materials and release hazardous substances into the environment. The rare-earth minerals in discarded electronics are rarely recovered efficiently, perpetuating the cycle of destructive mining.

Carbon emissions from AI operations continue to surge. Training a single large language model can generate as much carbon as five cars over their entire lifetimes. As AI models grow larger and more complex, their carbon footprint expands exponentially. The industry's promise of carbon neutrality often relies on carbon offsets rather than actual emission reductions, masking the true environmental impact.

Perhaps most troubling is the feedback loop between AI development and environmental degradation. As climate change intensifies, the demand for AI-powered solutions increases, driving greater resource extraction and energy consumption. This accelerates the very environmental problems AI is supposed to help solve. The mining of rare-earth minerals for "green" technologies like electric vehicles and renewable energy systems, many of which rely on AI, is causing irreversible damage to ecosystems that serve as crucial carbon sinks.

The environmental justice implications are stark. Resource extraction for AI predominantly occurs in the Global South, while the benefits accrue to wealthy nations and corporations. Communities that contribute least to AI development bear the greatest environmental burden, facing contaminated water supplies, destroyed agricultural land, and increased vulnerability to climate change.

Without immediate intervention, AI's environmental footprint threatens to undermine global climate goals and accelerate ecological collapse. The industry must transition from a model of unlimited growth to one of sustainable development, prioritizing efficiency, recycling, and genuine environmental stewardship over profit maximization.

Digital Infrastructure Control

Control over digital infrastructure has become a new domain of geopolitical competition, with nations vying for dominance over the physical and virtual networks that enable AI systems. Submarine cables, which carry 99% of international data traffic, have become strategic assets vulnerable to espionage, sabotage, and territorial disputes.

Recent incidents of cable cuts in critical locations have raised concerns about deliberate attacks on digital infrastructure. The concentration of cable landing points in a few geographic locations creates chokepoints that hostile actors could exploit. Nations are viewing these underwater networks as critical national security infrastructure, leading to new maritime tensions and surveillance operations.

The battle for satellite constellation dominance adds another layer to digital infrastructure conflicts. SpaceX's Starlink, Amazon's Kuiper, and Chinese alternatives are creating a new space race focused on global internet coverage. These constellations not only provide connectivity but also enable AI applications in remote sensing, military operations, and global surveillance. The dual-use nature of these technologies blurs the line between civilian and military applications.

5G and emerging 6G networks represent another infrastructure battleground. The exclusion of Huawei from Western 5G networks demonstrates how technological standards have become tools of geopolitical competition. As these networks become essential for AI applications, from autonomous vehicles to smart cities, control over network infrastructure translates directly into economic and military advantages.

Cloud computing sovereignty has emerged as a critical issue, with nations demanding data localization and domestic cloud infrastructure. The concentration of cloud services among a few American and Chinese companies has prompted other nations to develop sovereign cloud capabilities, fragmenting the global internet and creating new barriers to AI development and deployment.

Financial System Weaponization

The integration of AI into global financial systems has created new vulnerabilities and opportunities for economic warfare. AI-driven sanctions can now target specific individuals, companies, or sectors with unprecedented precision, while high-frequency trading algorithms can be weaponized to destabilize markets.

The exclusion of Russian banks from the global messaging network used by banks to coordinate payments, transfers, and settlements, known as SWIFT, demonstrated the power of financial infrastructure as a geopolitical weapon. AI systems that monitor and control these networks can implement sanctions instantaneously, but also create risks of algorithmic escalation where automated responses trigger unintended consequences. The development of alternative payment systems by China, Russia, and other nations reflects the growing concern over financial system weaponization.

Cryptocurrency and central bank digital currencies (CBDCs) represent a new frontier in financial warfare. AI systems that can track and analyze blockchain transactions threaten the anonymity of cryptocurrencies, while CBDCs could give governments unprecedented control over their citizens' financial activities. The competition among different digital currency systems could reshape global financial power structures.

High-frequency trading systems powered by AI have introduced new forms of market manipulation and flash crashes. These systems can execute thousands of trades per second, creating volatility that human traders cannot match. Nation-states could potentially use such systems to attack other countries' financial markets, creating economic instability as a form of hybrid warfare.

The concentration of financial AI capabilities in a few major financial centers creates new vulnerabilities. Cyberattacks on these systems could trigger global financial crises, while the algorithms themselves could be biased or manipulated to favor certain nations or economic blocs. The black-box nature of many AI trading systems makes it difficult to detect such manipulation until significant damage has been done.

Chapter 10

Critical Infrastructure Vulnerabilities

AI systems have become deeply embedded in critical infrastructure, from power grids to transportation networks, creating new vulnerabilities to cyberattacks and system failures. The integration of AI into infrastructure control systems offers efficiency gains but also introduces single points of failure that adversaries can exploit.

Power grids rely on AI for load balancing and predictive maintenance, but these same systems can be targeted by sophisticated cyberattacks. The 2015 attack on Ukraine's power grid demonstrated the vulnerability of computerized infrastructure. As AI systems gain more autonomous control over power distribution, the potential impact of successful attacks grows exponentially.

Transportation systems, including ports, airports, and rail networks, depend on AI for optimization and safety. Autonomous vehicles and ships introduce new attack surfaces that could be exploited to cause accidents or disrupt global supply chains. The NotPetya cyberattack's impact on Maersk shipping demonstrated how digital attacks can paralyze physical infrastructure.

Healthcare systems have become particularly vulnerable as AI integration accelerates. Ransomware attacks on hospitals have already caused deaths by disrupting critical care systems. As AI becomes more involved in diagnosis, treatment planning, and drug discovery, attacks on these systems could have catastrophic public health consequences.

Water treatment facilities, nuclear power plants, and other critical infrastructure increasingly use AI for monitoring and control. The Stuxnet attack on Iranian nuclear facilities showed how sophisticated malware can cause physical damage through digital means. As infrastructure becomes more automated, the potential for AI-enabled attacks to cause mass casualties increases.

Intellectual Property and Technology Transfer Battles

The race for AI supremacy has intensified intellectual property conflicts and technology transfer disputes. Industrial espionage targeting AI algorithms, training data, and hardware designs has become a major source of international tension. Nations accuse each other of stealing AI secrets while implementing stricter and stricter controls on technology exports.

Forced technology transfer requirements, where companies must share their AI technology to access certain markets, have become a major trade dispute issue. These requirements are seen as a way for countries to rapidly advance their AI capabilities at the expense of foreign companies. The practice has led to retaliatory measures and contributed to the decoupling of global technology supply chains.

The theft of AI training data and algorithms through cyberattacks has reached unprecedented levels. State-sponsored hacking groups target research institutions, tech companies, and government agencies to steal AI intellectual property. The value of these digital assets has made them prime targets for espionage, with some estimates suggesting billions of dollars in AI-related IP theft annually.

Patent wars over AI technologies have created a complex web of legal disputes that span multiple jurisdictions. Companies and countries race to patent AI innovations, creating thickets of intellectual property that can stifle innovation and create barriers to entry. The lack of international consensus on AI patentability adds another layer of complexity to these disputes.

The brain-drain competition for AI talent has become a form of intellectual property conflict. Countries use immigration policies, research funding, and other incentives to attract top AI researchers and engineers. This competition for human capital is particularly intense between the United States, China, and other major AI powers, with talent flows potentially determining future technological leadership.

Food Security and Agricultural Resources

AI's expansion is creating unexpected pressures on global food systems and agricultural resources. The conversion of agricultural land to data centers and tech campuses, combined with AI's water and energy demands, is impacting food production in several regions.

Industrial agriculture's increasing reliance on AI for precision farming, while potentially beneficial for efficiency, is creating new dependencies on technology that could be disrupted by cyberattacks or supply chain failures. The concentration of AI agricultural technology in a few companies creates vulnerabilities in the global food system.

Competition for arable land has intensified as tech companies seek locations for massive data centers. This is particularly acute in countries with limited land resources, where the choice between food production and technological infrastructure becomes a national security issue. The situation is exacerbated by climate change, which is already reducing the availability of productive agricultural land.

Water diversion from agriculture to tech infrastructure is creating tensions in water-scarce regions. In parts of India and the American West, farmers compete with data centers for water rights, leading to reduced crop yields and increased food insecurity. These conflicts are likely to intensify as both AI infrastructure and climate stress increase.

The use of AI in controlling global food supply chains has created new forms of food weaponization. Countries that control AI-driven logistics systems could potentially disrupt food supplies to adversary nations. The COVID-19 pandemic revealed the fragility of global food systems, and AI integration adds another layer of complexity and vulnerability.

Climate Migration and Resource Displacement

The intersection of AI development, climate change, and human migration is creating new patterns of conflict and displacement. As climate change renders some regions uninhabitable, the competition for resources in climate-resilient areas is intensifying, with AI infrastructure adding to these pressures.

Climate refugees are increasingly moving from vulnerable coastal and arid regions to areas suitable for AI development and tech industries. This creates competition for resources, housing, and employment in destination regions. AI-powered surveillance systems are being deployed to monitor and control these migration flows, raising human rights concerns.

The displacement of communities from resource extraction sites, particularly sites where rare-earth minerals are being mined for AI hardware, is accelerating. Indigenous populations and rural communities are often forced from their lands to make way for mining operations, creating environmental refugees who have lost their traditional livelihoods.

AI's role in climate prediction and adaptation could help manage migration flows, but it also enables more sophisticated border control and surveillance systems. The technology that could help identify climate-vulnerable populations is also being used to restrict their movement, creating ethical dilemmas about AI's role in climate justice.

Competition for climate-resilient locations suitable for AI infrastructure is driving up land prices and displacing local populations. Areas with stable climate conditions, renewable energy potential, and water resources are becoming premium locations for data centers, creating new forms of climate gentrification.

Arctic and Space Resources

The melting Arctic has opened new frontiers for resource extraction and AI-enabled exploitation. As ice coverage decreases, previously inaccessible mineral deposits, including rare-earth elements crucial for AI hardware, are becoming available for extraction. This has sparked a new "cold rush," with nations asserting territorial claims and deploying AI-powered systems for resource mapping and extraction.

AI-enabled autonomous systems are being developed for Arctic mining and drilling operations, capable of operating in extreme conditions with minimal human oversight. These technologies are accelerating the pace of Arctic exploitation, raising concerns about environmental damage to one of Earth's most fragile ecosystems. The military significance of the Arctic has also increased, with AI-powered surveillance and defense systems being deployed by Arctic nations.

The competition for space resources represents the next frontier in AI-driven resource conflicts. Lunar mining for Helium-3 and rare-earth elements, asteroid mining for platinum group metals, and the establishment of space-based manufacturing facilities all depend on advanced AI systems. Nations are developing legal frameworks to claim space resources, potentially setting the stage for the first conflicts beyond Earth.

Satellite constellations providing AI services, from Earth observation to communication networks, are creating new forms of space-based competition. Anti-satellite weapons and cyber capabilities targeting space infrastructure are proliferating, with AI systems both enabling these weapons and serving as targets. The militarization of space, driven partly by AI capabilities, threatens to extend earthly conflicts into orbit.

The dual-use nature of space technology, where civilian AI applications can quickly be converted to military purposes, complicates international cooperation. Nations are reluctant to share space technology that could provide adversaries with military advantages, leading to parallel development efforts and increased risk of conflict over orbital slots and frequencies.

Biological and Genetic Resources

AI's application to biotechnology has created new forms of resource competition centered on biological and genetic materials. The race to collect and analyze genetic data from diverse populations has led to accusations of biopiracy, where genetic resources from developing countries are exploited without fair compensation.

Biodiversity hotspots are becoming targets for AI-driven bioprospecting, where machine learning algorithms analyze vast databases of genetic information to identify potentially valuable compounds for pharmaceuticals, agriculture, and industrial applications. This has led to conflicts over access to protected areas and indigenous territories rich in biological diversity.

The control of agricultural genetic resources has become a critical issue, as AI enables more sophisticated crop breeding and genetic modification. Countries with diverse crop varieties and wild relatives of domesticated plants find themselves in disputes over access and benefit-sharing as AI companies seek to monopolize these genetic resources.

Digital sequence information (DSI) derived from genetic resources has become a contentious issue in international negotiations. AI systems can analyze this digital data without accessing physical specimens, potentially bypassing benefit-sharing agreements. This has led to calls for new international frameworks governing digital genetic resources.

The weaponization potential of AI-enhanced biotechnology adds a security dimension to biological resource conflicts. The ability to create targeted biological agents using AI-designed proteins or to disrupt agricultural systems through engineered pathogens has made biological resources a national security concern, leading to increased restrictions on international collaboration and data sharing.

Human Cost Estimates

The potential human cost of AI-driven conflicts is difficult to quantify, but we can model potential outcomes based on analysis of historical resource wars and geopolitical patterns. This can help us estimate potential death tolls under different conflict intensity scenarios.

Low-Intensity Conflict

In a Low-Intensity Conflict scenario centered on covert AI cyberwarfare and undermining economies, conflict primarily manifests through AI-driven cyberattacks targeting critical infrastructure, financial systems, and coordinated misinformation campaigns. Rather than conventional warfare, these operations focus on creating instability and disruption of essential services.

The human impact comes primarily from indirect effects: hospital system outages leading to patient deaths, transportation system failures causing accidents, and economic collapse triggering food insecurity and reduced access to medical care. While this form of conflict may appear bloodless on the surface, the estimated global death toll could reach between 100,000 and one million over a five- to ten-year period.

Mid-Intensity Conflict

A Mid-Intensity Conflict involving regional AI-enhanced wars would witness the deployment of AI-enhanced military technologies in regional disputes, including autonomous drones, battlefield automation systems, and advanced AI surveillance networks. Potential flashpoints include escalation in the Taiwan Strait, the India-China border, or across Middle East conflict zones.

The human cost would include direct military and civilian casualties from automated or AI-targeted strikes, as well as deaths from displaced populations and infrastructure collapse. Each major regional conflict employing these technologies could result in an estimated five to 20 million deaths, significantly higher than comparable conventional conflicts due to AI's ability to accelerate targeting, reduce decision time, and potentially bypass traditional defenses.

High-Intensity Global Conflict

The most severe scenario, a High-Intensity Global Conflict or AI-driven world war, involves full-scale war between major powers such as the United States, China, and Russia, employing fully autonomous weapons systems, space-based AI platforms, and AI integration into nuclear command and control structures. The dangers in this scenario come from several factors: rapid escalation due to AI misjudgments or errors, unrestrained cyber-kinetic warfare targeting civilian infrastructure, and the potential use of weapons of mass destruction either controlled by or defended with AI systems.

The projected human toll is catastrophic: 50 to 100 million deaths in the first months of conflict alone. Long-term consequences, including potential nuclear winter and societal collapse, could result in deaths numbering in the hundreds of millions to over one billion people worldwide.

Additional Factors

Several additional factors could significantly influence these projections. The degree of autonomy versus human oversight maintained over AI weapons systems will dramatically affect casualty rates. Fully autonomous systems without meaningful human oversight present the highest risk of accidental escalation and unintended casualties. The combination of AI with biological technologies creates particularly severe risks. AI-enhanced bioweapons could potentially trigger pandemics with death tolls exceeding several million people globally.

The development and implementation of robust control and ethics frameworks, including international treaties and AI governance mechanisms, could significantly reduce these risks. Nations that establish clear red lines around AI weaponization and maintain human control over critical military systems may help prevent the worst-case scenarios outlined above.

These projections illustrate the potential scale of devastation if AI-driven resource conflicts continue to escalate unchecked. Even under the lowest-intensity scenario, over a million deaths could occur, highlighting the growing risk of AI-induced geopolitical competition.

Ethical Imperative

The expansion of AI is not just a technological issue. It is a moral and ethical dilemma with far-reaching consequences. The unchecked growth of AI-driven industries is intensifying demand for critical resources, escalating conflicts and deepening global inequalities.

Microsoft CEO Satya Nadella has warned of an "overbuild" of AI infrastructure, suggesting that the current pace of AI development may be excessive. He noted that future AI capacity might be leased rather than owned, implying that the rapid expansion of AI systems could lead to resource inefficiencies and unintended consequences.

While AI has transformative potential in fields such as medicine, security, and infrastructure, a significant portion of AI computing power is currently consumed by entertainment applications, AI-generated influencers, and excessive commercial use. Redirecting AI development toward meaningful and socially beneficial applications could alleviate pressure on global resources and reduce the risk of resource-driven conflicts.

One potential solution is the implementation of an AI Resource Tax, which would discourage wasteful AI applications by taxing models based on their consumption of rare-earth minerals and energy. Such a measure could incentivize efficient AI development, ensuring that computing power is allocated to critical fields rather than excessive profit-driven ventures.

Strict oversight of AI data centers and resource allocation is essential to ensuring that AI serves human progress rather than corporate profit or military dominance. Without thoughtful governance, the race for AI supremacy will continue fueling conflicts, deepening economic divides, and exacerbating global inequalities.

Balancing Progress

The rapid growth of AI is a geopolitical transformation with profound consequences. As nations compete to secure the critical resources essential for AI development, conflicts over rare-earth minerals and energy supplies are intensifying. Without strategic oversight, the unchecked expansion of AI threatens to destabilize entire regions, fueling economic warfare, resource-driven conflicts, and military confrontations that could claim millions of lives in the coming decades.

The integration of AI into military strategy and economic competition has added a new dimension to global conflicts. AI-powered automation in defense, surveillance, and cyberwarfare increases the risk of rapid escalation, as autonomous systems make split-second decisions that could remove traditional diplomatic buffers.

Proactive measures must be taken to develop international frameworks that ensure AI technologies are used responsibly. Governments, corporations, and global institutions must collaborate to create regulations that limit AI's potential to fuel geopolitical instability. Without such safeguards, AI could exacerbate economic disparities, accelerate the depletion of critical resources, and serve as a catalyst for the deadliest conflicts of the 21st century.

"I know not with what weapons World War III will be fought, but World War IV will be fought with sticks and stones." — Albert Einstein

Make no mistake, AI is a weapon of war. We should be treating it as nothing less.

Chapter 11
The Choice Before Us

The Moment When Everything Is Still Possible

Human history has always had pivotal moments of choice, when multiple paths stretched forward and the outcome was undetermined. The choice of path represents untapped possibilities waiting for human decision to bring one reality or another into being. The great transitions in the history of mankind have demonstrated that the trajectory of human progress is not by chance but through deliberate human choice and action. Transformations required struggle and sacrifice, emerging through collective will rather than predetermined outcomes.

We stand at such a crossroads now, though it manifests not as a single dramatic moment but as a pervasive tension throughout society. This critical point appears in the anxiety of displaced workers confronting automation, in parents' uncertainty about their children's future, and in young people questioning their role in an automated world. Communities everywhere experience the paradox of systems that can predict consumer behavior with remarkable accuracy while failing to address deeper human needs and aspirations.

The well-traveled path before us leads toward a future where techno-logical systems grow ever more intelligent while human connections weaken. On this trajectory, efficiency metrics improve as meaningful work diminishes, freedom becomes segregated by economic access, and dignity becomes secondary to optimization. As we follow this path, dissatisfaction grows in predictable patterns, visible in our social data yet unaddressed by our systems.

Yet, alternative paths remain available to us. We could choose delib-erate restraint to examine the human implications of technological progress. We could design systems that enhance rather than diminish human capabilities, build community structures that prioritize connec-tion over convenience, and develop economic models that serve us beyond mere technological advancement.

The opportunity to prioritize human considerations remains viable, but technological momentum creates urgency. Each day, automated systems assume more decision-making authority, and each month brings new technologies that replace human judgment with algo-rithmic processes. The gravitational pull toward dehumanization intensifies with every innovation we adopt without examining its human impact.

The uncomfortable truth is that institutional systems, market forces, and technology itself cannot generate human-centered solutions autonomously. Only deliberate action, grounded in memory of what matters most, can preserve the essential human elements that give life meaning and value.

This Is Only the Beginning

From my decades of observing technological transformation, with all its elevating innovations and isolating developments, I want to offer this invitation to maintain awareness of the fundamental human quali-ties that define our species.

These qualities extend far beyond computational capacity or productivity metrics, beyond any ability to compete with machines. They encompass our capacity for empathy, our drive to create beauty without functional purpose, our ability to be present with both suffering and joy, and our unique human talent for finding meaning in patterns that algorithms can only process but never understand. This acknowledges the inherent uncertainty in assessing technology, and recognizes that we often embrace innovations we later question, and question developments we later embrace.

Through this ongoing process of evaluation and re-evaluation emerges a deeper commitment to defending essential human qualities. This defense requires vigor, creativity, and clarity because these qualities face erosion, not through dramatic collapse but through the gradual diminishment of dignity, meaning, and connection. The danger lies in incremental loss, until one day we can no longer recognize ourselves within the systems we've created.

The Fork in the Road

Two primary trajectories stretch before us: one leading toward a machine-centered future, the other toward a human-centered alternative. This choice will shape not merely our economic and political structures but our fundamental conception of what it means to be human in the age of artificial intelligence.

The Machine-Centered Future

In the machine-centered trajectory, efficiency shifts from a helpful metric to a moral conviction. Corporations treat optimization as inherently virtuous and justify any human cost in its pursuit. Data supersedes human wisdom, with quantitative measures overriding human experience and knowledge. Systems like these prioritize metrics over lived experience, treating human observation as less valid than numerical indicators.

Prediction gains priority, and control appears more practical than trust. Recommendation systems, such as those built into social media platforms, demonstrate this pattern perfectly. They do this by narrowing our exposure to the world while claiming to expand our horizons, reinforcing existing patterns rather than introducing perspectives that might challenge or transform us.

This trajectory creates a world where human value depends entirely on system utility, where people are measured solely by their ability to add value to autonomous systems. Those unable to match the technological pace face marginalization: elderly populations, neurodivergent individuals, and anyone whose skills fall out of alignment with current market demands.

Community structures become purely transactional, with digital convenience replacing human interaction at every turn. Meaning itself becomes a commodity that transforms life experiences into products for consumption rather than opportunities for authentic engagement. A world like this risks internal collapse through the gradual suffocation of those elements that give life substance and significance.

A Human-Centered Future

Contrary to the machine-centered future, a human-centered trajectory offers a much different vision, where efficiency remains relevant without becoming obsessive. This way forward doesn't suggest abandoning technological progress or retreating to the Stone Age. Instead, it promotes efficiency as a tool that helps humans thrive rather than replacing them. In this future, creativity, purpose, and worth receive protection as inherent values, not luxuries. Organizations emerge that measure success through their ability to enable meaningful lives alongside productive output.

This trajectory still includes struggle. It offers no utopian fantasy. However, struggles are faced collectively, with central questions focused on how to live well together rather than how to win against one another.

The Future Is Not Inevitable; It Is Shaped by Our Choices

The belief in technological determinism — that innovation follows an inevitable path we must passively accept — lacks any real foundation. The history of technology reveals multiple decision points and alternative outcomes at every stage of development. We do have choices — but do we possess the courage to make them? Each decision we make represents an opportunity to shape technology according to human values rather than allowing technology to reshape humanity.

The Moral Crossroads

At the center of our current choice lies a moral rather than a merely technical question. This decision concerns our identity as a species, what we're willing to sacrifice, and what we're determined to protect. The crossroads we face demands clear moral reasoning about fundamental values.

Profit vs. Purpose

We confront priority decisions that pit short-term gains against long-term meaning, shareholder value against human well-being, and growth metrics against the health of communities, families, and ecosystems. Markets, despite their utility, lack moral neutrality. Every economic choice carries human consequences, whether we acknowledge them or not. Will we treat people as means to our ends or as ends in themselves? This ancient philosophical distinction takes on new urgency in an age when algorithms can optimize for any goal we set, making the choice of goals more critical than ever.

Control vs. Creativity

Today's controlled processes for system design force us to choose between managing and predicting behavior or enabling human imagination. We must decide whether to optimize for compliance with a process or to allow for discovery and surprise. Control promises predictable outcomes and safety, which offers us the comfort of known outcomes and manageable risks. Creativity, conversely, requires trust and tolerance for uncertainty, demanding that we accept messiness as the price of innovation. We cannot fully possess both qualities; we must choose which we value more and design our systems accordingly.

Humans as Resources vs. Irreplaceable Ends

Perhaps the deepest choice concerns how we conceptualize human beings themselves. Do we view people as extractable resources or as lives deserving of honor? Are individuals merely data points to be aggregated or unique beings whose experiences carry irreplaceable meaning? Various populations, including elderly care recipients, struggling students, refugees, and unprofitable artists prompt this essential question: Are they inefficiencies requiring solutions, or mirrors reflecting our shared humanity? How we answer will define not just our technologies but the character of our civilization.

Why This Crossroads Matters

Current generations face a technological power of unprecedented scope, pervasiveness, and intimacy. We are the first to grapple with artificial intelligence that can mimic human conversation, predict human behavior, and increasingly replace human judgment. This puts us in a unique position as the generation that will determine what it means to be human in an age of machines, a responsibility that cannot be deferred or delegated.

Signs of Hope

Despite the magnitude of these challenges, pessimistic surrender ignores widespread evidence of resistance, renewal, and human possibility. Around the world, persistent signals of hope emerge from diverse sources, suggesting that alternative futures remain achievable.

Ethical AI Initiatives: Innovation Aligned with Conscience

Across research institutions, corporations, and policy circles, dedicated professionals work to embed ethics within innovation itself. Many work in silence to ensure that transparency, fairness, and accountability are embedded in the work they do. Though these efforts remain imperfect and often underfunded, they demonstrate that, when we choose to unite ethics and innovation, they can work as partners rather than opponents.

Local Resilience Projects: Small-Scale, Human-Centered Change

Communities worldwide are building solid alternatives to dominant technological paradigms. Local food networks restore relationships among people, land, and neighbors. Cooperative businesses prioritize collective well-being over extractive profit models. Digital commons resist surveillance capitalism while rebuilding public trust in shared resources. These projects rarely make headlines, yet they quietly demonstrate that bottom-up change remains possible through patient, relationship-by-relationship, choice-by-choice development.

Cultural Shifts Toward Meaning: The Hunger for More Than Optimization

Even within highly digitalized societies, growing numbers of people express a preference for deliberation over speed, depth over distraction, and purpose over mere performance. Movements addressing mindfulness, mental health, work-life balance, social justice, artistic expression, and ethical technology reveal a shared longing for wholeness that efficiency alone cannot provide. This cultural restlessness, far from representing weakness, creates fertile ground for transformative change.

Why This Hope Matters

Hope transcends naive optimism or denial of difficulty. True hope represents a refusal to accept inevitable futures, embodying the conviction that humans retain the power to choose differently, build alternatives, and begin anew despite overwhelming systemic pressures. Throughout history, hope has driven every meaningful change, and we need it now more than ever.

The Weight of Our Time

Each generation faces its defining challenges, and ours is being tested not by a single crisis but by a pervasive question about human meaning in machine contexts. This question is simultaneously philosophical and intensely practical, manifesting in countless daily decisions about tool design, organizational structures, educational approaches, urban planning, legal frameworks, and interpersonal interactions.

Future generations will judge us not by our algorithmic sophistication, market efficiency, or device aesthetics. Historical, descendant, and conscience-based judgment will center on whether we remembered the humans, and whether we preserved and protected what makes us fundamentally human in the age of artificial intelligence. Technological immersion provides perspective on both triumphs and unintended consequences, generating varied optimism and pessimism across time. Yet one recognition remains essential: Surrendering the future is not necessary.

We retain the ability to shape what comes next —to bend the arc of technological development toward justice, beauty, belonging, and authentic human fulfillment. The declaration of success for humans requires recognizing the significance of our historical moment and rising to meet its challenges with wisdom, courage, and unwavering commitment to human dignity. The choice presents itself clearly. The moment is now. The future remains open to those willing to claim it.

Part Four
The Path Forward

Chapter 12
Reasons for Hope

Hope Is Not Foolish

After several chapters that document how advances in technology can forget or even hurt us, you might be wondering if resistance is futile. I don't believe it is and I want to show you why it's not. There is hope. Come with me and let's explore some real examples of hope in action.

Organizations and communities are making choices that are beneficial to everyone. It's not all doom and gloom. Some people get it and are putting their beliefs into practice.

No organization gets everything right, and I don't want you to come away thinking I believe that. Organizations are made of people who have varied belief systems. It's not logical to think they are going to do everything right in all divisions of the organization. However, progress at scale is still possible, and there are some great examples we can explore together. These examples aren't just theoretical; they are being implemented successfully right now.

Negative Outcomes Are Not Inevitable

Before we get to specific examples, I need to address something important. You might feel that there is no hope after reading about the machines winning from Part I, or the devastating human costs in Part II, and the moral collapse warned about in Part III. I want to assure you that, no matter how bleak they seem, those outcomes are not inevitable. We can still reverse course in some areas to ensure that hope is not extinguished.

Organizations exist that, despite facing the same pressures as every other organization, have chosen a human-centered path. They decided to dispel the myth that dehumanization is the inevitable price for progress and profit in their industry. In fact, they may see something that their competitors do not see. Human-centered choices aren't only ethical; sometimes they are profitable. Sometimes they can give you a competitive advantage.

Examples of Human-Centric Choices

No company is perfect and, although some have exhibited the qualities promoted in this book, not all do so consistently. At the time of this writing, the organizations discussed below exhibit the qualities I have extolled.

Preserving Human Connection

When the pressure of automation through self-checkout kiosks began, many retailers rushed to implement them and reduce the headcount of their cashiers. But popular grocery chain Trader Joe's decided to take a different approach. They realized the strength of their brand was powered by their employees providing a personalized customer experience.

Walking into a Trader Joe's gives me a feeling like no other grocery store. The "crew members," as they are called, are approachable, knowledgeable, and give personal advice. I am on a low-carb diet and constantly on the hunt for low-carb lunch solutions. On a visit to Trader Joe's, I asked a crew member for help finding a low-carb lunch.

This young lady went out of her way to help me. She recalled a recipe she saw on a social media platform and walked me through piecing together an amazing Tuna and Guacamole wrap from products in the store. She checked the ingredients to ensure I would be within my carb limit and didn't let me forget to try a Blueberry Lemon Sparkling water. I don't think I would have had the same experience at any other grocery store. I hesitate to even call it a grocery store. It's something more than that, and I think that is exactly what the leaders want us to feel.

It gets better, though. The checkout was a fast but personal experience. I felt like they genuinely cared about me. I don't think it was fake either; that feeling permeated throughout the entire store.

Why wouldn't it? When faced with outsourcing jobs to automated kiosks, Trader Joe's now-retired President, Jon Baseline, doubled down on not getting rid of crew members regardless of any perceived efficiency gains. The company believes in their people, and that support ensures they preserve the personal and friendly customer service experience their competitors can't provide.

The brilliance of the leadership to stop and question what their competition was doing has set them apart. In a world where many retail chains are shuttering locations, Trader Joe's plans to continue expanding.

If we need a poster child for the principles in this book, look no further than this company. When employees feel valued, customers experience an authentic connection. This is a testament to brand loyalty through human dignity. No amount of technology can replace that human-to-human experience and the loyalty it produces.

Mastering the Machines

Arguably one of the most reliable car brands on the road, Toyota has a reputation built on quality and dependability. During the early 2000s, pressure increased in the automotive industry to eliminate human inefficiencies. Many of their competitors pushed to automate and lay off employees. When the financial crisis of 2008 to 2010 occurred, their competitors laid off significantly more of their staff in favor of fully automating those jobs.

While this was going on, Toyota faced several other challenges. They had a significant reputation crisis due to some large-scale recalls that included quality issues. Their *kaizen* (continuous improvement) culture was starting to erode, and the younger generation of workers was becoming increasingly detached from the physical manufacturing process due to the heavy automation Toyota had implemented. Craftsmanship was slowly dying as institutional knowledge left the building and was replaced by humans who did not know the underlying craft operating machines.

In response to these challenges, Mitsuru Kawai, a senior technical executive, championed a different approach. He reintroduced skilled human labor into processes that had been automated. This strategy brought back one of Toyota's core principles of *jidoka* — automation with a human touch. By putting the robots aside, he had workers manually produce parts. Workers became machine teachers instead of machine babysitters. This approach resulted in the workers gaining a deeper understanding of the processes. The expertise they gained allowed them to identify process and quality improvements that might have been missed by the automated machines.

Toyota reported that these teams focused on quality were able to generate ideas that not only produced better quality but improved productivity. One production line is said to have been shortened by 96% and additionally reduced material waste by 10%. The payoff went well beyond higher-quality output — it saved jobs and empowered employees.

This is yet another example of how investing in human expertise can boost efficiency and innovation with less negative impact than implementing technology such as automation just to keep up with industry trends.

Designing Work for an Aging Workforce

In 2007, BMW recognized that the age of their workers on the production line would increase significantly in the next ten years. They had a choice to make. They could follow the trends of many automakers at the time by heavily automating the jobs and pushing older workers out. Instead, the leaders of BMW decided to experiment with making the production lines more age inclusive.

A pilot project called "Today for Tomorrow" was conducted on one assembly line, involving several workers with an average age of 47. This was planned to simulate the demographic conditions expected on production lines in the future. The employees were able to use their expertise and experience to implement several low-cost changes to the line. These changes included adjustments to tool positions, better lighting, and improved flooring to reduce joint stress. In total, over 70 changes were introduced. These resulted in a 7% increase in productivity, a zero-defect rate in production, reduced absenteeism, and overall higher job satisfaction. The project was so successful that BMW rolled it out to their plants around the world.

The company could have easily taken the route of planning the replacement of their aging and loyal workers. Many companies make that decision every day. Instead, BMW was looking out for their employees a decade down the road. They realized the retention of valuable employees would bring more to the quality and efficiency of the product than heavy automation would. This is yet another example of an organization winning by focusing on human-centered improvements while still getting performance gains with technology.

Real Change Is Realistic

These are just three examples of organizations that, in the not-so-distant past, have considered the consequences of a world where humans are not taken into account. They are examples of what happens when leadership within organizations acts with courage and conscience. Human-centered choices are possible at every level, from the individual to the organizational to the societal.

This ability to choose humanity gives me hope. The warnings in this book do not need to become prophecies. Every person in these organizations who stood up and challenged the narrative proves that the future remains unwritten.

In the next part of this book, we'll explore how we can implement these principles. Each one of us has a sphere of influence where we can plant the seeds of these principles so they can take root around us. Negative outcomes are not inevitable. The future is still in our hands. Don't lose hope.

Chapter 13
Building Human-Centered Systems

If We Don't Shape Technology, It Shapes Us

Technology is never neutral. Each system we create either strengthens human autonomy, dignity, and connection, or it erodes them in the name of efficiency, profit, or control. This isn't philosophy; it's observable reality across industries, organizations, and communities. Every government database, education platform, hiring algorithm, or social media feed we build makes a choice about human worth.

When workforce management systems track productivity down to the keystroke and flag "inefficient" employees, they're not just measuring work; they're redefining it. The executives who applaud the metrics and dashboards often miss what the team leads quietly recognize: These systems change the nature of work itself, and in that transformation, something essential about human judgment and agency is lost.

No Neutrality in Systems

If technology is not designed for humans, it will eventually be designed against them. Every input reflects a choice. Every metric reflects a value. Every outcome shapes the world real people must live in. We talk about the future as if it's something mysterious that will arrive and reshape us, but the truth is both more sobering and more empowering: The future isn't happening to us; we're building it right now, decision by decision, system by system.

Designing for the Whole Human Being

Like a well-made tool that fits the hand perfectly, systems should adapt to people rather than forcing people to adapt to them. The question before us isn't whether we will shape technology, but whether we will have the courage to design technology that serves the whole human being: not just the market, not just the machine, not just the metrics.

This work begins with fundamental priorities: We need autonomy over dependency, creating systems that expand human agency rather than replace it. We need meaning over engagement, designing for depth of purpose rather than superficial stickiness. We need connection over convenience, building technology that sustains real relationships instead of just enabling faster interactions. And we need dignity over data extraction, creating platforms that protect humanity rather than monetize its vulnerabilities.

These aren't lofty ideals — they're survival requirements for the coming age. This shift cannot be left only to engineers or executives. It demands a new generation of leaders, designers, educators, and citizens who will not trade human dignity for speed, who will not accept the slow dehumanization of systems as "progress," and who will insist that we remember the humans at every level, in every design, in every decision.

Principles for Human-Centered Design

Engineering teams often focus on maximizing throughput or efficiency without considering what that means for the people using their systems. Human-centered design isn't just about aesthetics or user experience buzzwords: It's the disciplined, values-driven commitment to create systems that prioritize people in their complexity and variability above performance alone.

Dignity

When warehouse workers are guided by algorithms that dictate their every move — which item to pick, which path to take, how long each task should take — they often feel reduced to mere extensions of the system. Experienced workers who have developed expertise over years find their knowledge dismissed by algorithms that assume they need constant direction. Systems that degrade dignity by reducing people to output, engagement, or compliance may succeed in the short term, but they fail where it matters most: in building loyalty, trust, sustainability, and care.

Human-centered systems protect space for rest, growth, and even dissent. They recognize the inherent worth of the individual beyond their utility to the system. Manufacturing companies that implement process management software without preserving employee discretion often lose their most experienced workers; not because of inadequate compensation, but because the work no longer allows them to apply their accumulated knowledge and judgment.

Transparency

Automated hiring systems that reject qualified candidates without explanation create frustration that goes beyond the rejection itself — it's the opacity of the process that breaks trust. When decisions affecting jobs, freedom, or access happen behind black-box algorithms, and people can't understand why or how to challenge them, the relationship between institutions and individuals fundamentally erodes.

Transparency is more than visibility — it's about accountability. People must be able to understand, question, and engage with the systems that shape their lives. When hospitals make their patient scheduling algorithms transparent to both staff and patients, complaints often drop dramatically. This isn't because the system suddenly becomes perfect, but because people can see it working, understand its constraints, and suggest improvements. Visibility creates voice, and voice creates both better systems and greater buy-in.

Flexibility

Machines thrive on standardization, but humans do not. Financial-aid processing systems designed for maximum efficiency through strict categorization often fail when faced with the complexity of real students' lives: blended families, unusual financial situations, and life disruptions that don't fit neat boxes. Real people are messy, diverse, and unpredictable, bringing different histories, abilities, contexts, and needs.

Systems must flex to meet people where they are rather than penalizing them for failing to conform. The most resilient systems build in space for exceptions, overrides, and human judgment. They treat variability not as a bug to be eliminated, but as a fundamental feature of authentic human life. This isn't inefficiency, it's wisdom.

Challenges in Implementation

Agreeing that systems should be human-centered is one thing; building them that way is another. Even well-intentioned leaders struggle to bridge the gap between principle and practice. Several core challenges stand in the way of making human-centered systems real.

Complexity

Automated decision systems for child-welfare cases exemplify how complexity becomes a barrier to human understanding. As variables multiply, decision trees branch, and data points accumulate, systems can become too complex for any single person to fully grasp or audit. At what point do we lose the ability to determine whether these systems are helping or hurting the real children they're meant to serve?

As systems become more complex, particularly those powered by AI, it becomes exponentially harder to keep people meaningfully involved.

Algorithms make decisions in milliseconds that no human can audit in real time. Interconnected systems create ripple effects no single designer can fully foresee. Decision-making becomes "too fast to question" precisely when questioning is most needed. This isn't just a technical problem, it's also a governance problem that forces us to ask who has the right and ability to pause the system when it starts hurting people.

Economic Pressure

The tension between doing what's profitable and doing what's right has always existed, but the speed and scale of technological change has intensified this dilemma. In competitive environments, ethical concerns often meet resistance: "If we don't do this, our competitors will." We operate in markets where speed and profit are rewarded, often at the direct expense of well-being.

Features that keep users scrolling are seen as successes, even when they worsen mental health. Hiring algorithms that filter applicants more quickly are applauded, even when they systematically exclude whole populations. Ethical design often costs more and takes more time, raising hard questions that investors would rather not face. In this environment, moral courage can feel like a liability, and good people make compromises they later regret because the pressure to deliver results overwhelms their deeper values.

Cultural Inertia

Organizations where individuals understand the need for human-centered systems often find the institution itself resists change. "That's not how we do things here" becomes a shield against improvement. Institutions are not designed to adapt easily: Bureaucracies stagnate, protocols become sacred, and established practices become barriers to necessary change.

Even when individuals inside organizations care deeply, they're often fighting against structures built decades ago: structures that assume conformity, predictability, and control. Perhaps most challenging is when people become loyal to systems that hurt them because it's the only way they know. Social workers using broken case management systems that force hours of data entry instead of client interaction might resist proposed improvements, saying, "At least we know how this one works." The fear of disruption becomes stronger than the hope for something better. Changing this requires more than better tools. It demands vision, persistence, and often a sustained struggle.

Strategies for Change

Naming what's wrong isn't enough. We need practical roadmaps and models. Across industries and institutions, some people are already doing this work — not perfectly, but persistently.

Human-in-the-Loop Design

One of the most dangerous myths of the digital age is that full automation is always better. School districts implementing early warning systems for students at risk of dropping out demonstrate this principle. When systems flag students based solely on algorithms analyzing attendance, grades, and disciplinary records, they miss crucial context that teachers possess about their students as whole people. When redesigned to support rather than supplant teacher judgment, these systems transform into tools that amplify educators' capacity to care rather than replacing their judgment with mechanical rules.

In healthcare, AI diagnostic tools assist radiologists without supplanting them. The system flags potential anomalies, but trained doctors make the final call. This creates a collaborative loop combining machine speed and precision with human context. This approach preserves professional merit, invites judgment, and protects against blind trust in machine decisions. The human remains meaningfully in the loop.

Participatory Governance

When systems impact the public, the public should help shape them. This becomes especially critical for systems like predictive policing algorithms in communities that have experienced biased policing for generations. When these systems are developed without input from affected residents, they risk perpetuating historical injustices under the guise of objective technology.

Participatory governance isn't about endless debate; it's about inviting lived experience into design and decision-making. Cities that invite citizens to co-develop data privacy policies by asking residents what kind of data feels intrusive, how they want to be notified, and what opt-outs should exist, create policies that may not be perfect but are profoundly more human. This approach honors local knowledge, builds trust, and generates policies people understand and believe in.

Ethical Audits

Before launching new features or policies, software companies routinely check for bugs, security flaws, and financial risks, but systematic checks for ethical harm remain rare. Automated benefit determination systems that leave vulnerable families without access to critical services often pass all technical functionality tests while failing to examine impacts on the most vulnerable users.

Ethical audits are emerging as critical tools. They serve as structured ways to evaluate how systems affect human dignity, autonomy, and inclusion. Some tech firms now conduct "red team" exercises where experts, including ethicists and affected community members, stress-test systems for unintended consequences. This approach reveals harm early, embeds ethics into development rather than relegating it to public relations, and forces uncomfortable but necessary questions before it's too late.

Understanding System Success and Failure

Medical staff scheduling tools provide instructive examples of how systems can fail or succeed based on their approach to human needs. Systems that prioritize pure optimization, maximizing shift coverage and minimizing costs, often create technically flawless but humanly broken solutions. Nurses scheduled for impossible shifts, new physicians rarely paired with experienced mentors, and effective teams constantly shuffled all represent the human cost of narrowly defined efficiency.

When systems are redesigned to preserve team cohesion, protect rest time, and prioritize mentoring relationships — with built-in overrides allowing floor managers to adjust based on human needs that algorithms can't see — the results transform. Not only do satisfaction and retention improve, but the systems become more effective overall. Medical errors decrease, patient satisfaction increases, and organizations often save more money in the long run through improved staff retention.

This illustrates a fundamental principle: Systems succeed when they serve people, not when people are reshaped to serve systems. When people must contort themselves to fit rigid systems, they burn out, disengage, or break. But when systems flex around the needs of real humans, creativity returns, loyalty deepens, and integrity becomes possible again. The best systems do more than function — they honor.

Honoring Human Needs in the Post-Information Society

At the intersection of technological capability and human experience lies our most critical choice. We have built systems of unprecedented power, yet for all their sophistication, these systems cannot determine their own purpose. That remains uniquely our responsibility.

The technologies we've created have outpaced our conversations about why we create them. We've focused intensely on what machines can do, while neglecting deeper questions about what they should do, and more importantly, what they should never do.

This imbalance between capability and purpose creates the central challenge of our era: How do we harness technological advancement while ensuring it serves rather than subverts human needs? How do we create systems that amplify our humanity rather than diminishing it?

The answer begins with clarity about what makes human experience valuable beyond efficiency or productivity. It requires identifying which aspects of being human must be protected even when they seem inefficient by algorithmic standards. And it demands the courage to draw boundaries that may slow advancement but preserve meaning.

This is not a technical challenge but a moral one. It asks us to become stewards of both progress and preservation, to hold seemingly opposing values in creative tension rather than sacrificing one for the other.

The Post-Information Society Will Survive Only If It Honors, Not Erases, Human Needs

Here is the great paradox of our time: The more powerful our machines become, the more urgently we must protect the needs only humans carry. If we forget autonomy, mastery, purpose, belonging, and dignity, we will build a society that may be technologically dazzling, but spiritually bankrupt.

But if we remember and design systems that serve and elevate these needs, we have the chance to build something astonishingly rare: a world where progress and humanity move forward together.

As E.F. Schumacher wisely observed, "Any intelligent fool can make things bigger, more complex, and more violent. It takes a touch of genius, and a lot of courage, to move in the opposite direction."

We need that touch of genius now — the courage to make technology more human, not humans more like technology.

The Systems We Build Reflect What We Value

The systems we build are not just tools. They are mirrors reflecting what we value, what we prioritize, what we are willing to sacrifice, and what we are determined to protect. Every algorithm encodes a choice. Every policy embeds an ethic. Every design decision carries implications about who we believe people are and what they deserve.

When we build systems that treat humans as data points, we create environments where people feel disposable. When we optimize for clicks, we erode meaning. When we confuse engagement with belonging, we deepen loneliness. Corporations wonder why innovation stalls, loyalty fades, and something essential seems to slip away, not recognizing that their systems are suffocating the human spirit they depend on.

Business leaders must ask themselves a critical question: How are your customers truly responding to the technology you've implemented? Beyond metrics and satisfaction surveys lies a deeper truth that requires an honest assessment. Are your customers genuinely embracing these systems, or are they merely tolerating them because alternatives don't exist? Look carefully for the early warning signs, the subtle friction points, the workarounds users create, and the reluctant compliance that masks growing frustration. These seemingly minor issues may represent the first drops from a leak that, left unaddressed, will eventually compromise your entire foundation.

I believe AI is going to drastically amplify this dynamic, creating unprecedented disconnects between what organizations think they're providing and how humans experience these systems. The question isn't whether your technology functions as designed, but whether it serves authentic human needs in ways that strengthen rather than erode relationships. This isn't just about customer retention; it's about whether the systems you've built honor or diminish the people they're meant to serve. It's important to ask this question now before the damage becomes irreparable.

Chapter 13

Systems can also reflect our best aspirations. They can mirror our courage, carry forward our care, and amplify our deepest commitments to dignity, creativity, belonging, and meaning. Organizations that design systems with profound respect for human capability and complexity see the difference in their people's energy, their willingness to bring their full selves to work, and their capacity for innovation.

Chapter 14
Navigating the Post-Information Society

What Happens When We No Longer Control Information

T he sound of a dial-up modem handshake is something I can never forget. The seemingly disorganized pattern of screeching and hissing sounds was commonly heard in the early days of the internet. That analog sound represented access to more information than I could possibly have imagined. The ability to access libraries of information and entire encyclopedias from a keyboard and a screen was intoxicating. I spent hours tying up my family home's landline, accessing and downloading information. This was a new frontier to explore, and I was determined to learn every-thing I could about it. I was in control of my own destiny, but more importantly, I was in control of vast amounts of information.

It was called the digital revolution, and it was sold to us with the promise that information would liberate us. I was a true believer. In the 1990s, personal computers and the internet opened unprecedented access to knowledge and communication to ordinary people, offering pure potential by providing us with tools that would empower us beyond the traditional constraints of wealth, privilege, or gatekeepers who controlled access to knowledge.

This new technology appeared to deliver on this promise as well. The communication tools provided by the internet broke down barriers between communities and cultures. Average citizens gained access to information that enabled them to question authority and organize movements. Knowledge was able to spread through networks rather than hierarchies.

Today's information environment presents different challenges. Rather than facing a scarcity of information, people are dealing with an over-abundance of it, much of it generated, filtered, or manipulated by artificial intelligence systems and malicious actors. The distinction between authentic and synthetic content has become increasingly difficult to recognize.

The term Post-Information Society doesn't mean that information has disappeared; it's actually the opposite. We're drowning in a sea of content. The critical change lies in our relationship to unmediated reality and verifiable truth. Algorithmic systems currently govern the majority of information flows, determining which content is disseminated to which audiences, thereby shaping narratives.

With the mountain of digital clutter we have now, it's getting harder to tell what's real and what's fake. Machines are becoming more and more sophisticated and adaptable to behavioral patterns, and it's getting harder to tell what's human and what's not, what's true and what's false. Our ability to comprehend and navigate this evolving environment requires constant effort to preserve human agency and discernment.

The Post-Information Society presents a fundamental challenge: preserving human capacities in an environment increasingly shaped by machine intelligence.

The Map of the New World

The pace of technological change can feel disorienting for many people. Some people may experience a sense of displacement as once familiar patterns of information consumption and social interaction rapidly change around them. This can already be seen, for example you can't go far without bumping into something that claims to have AI integrated.

The past strategy of staying informed and keeping up with news is no longer viable for navigating the modern world. The challenge has shifted from accessing information to evaluating its nature and authenticity. We now operate in a world where reality blends with simulation, facts and deception are intertwined, and genuine human presence competes with digital performance.

The Post-Information Society demands different forms of adaptation than previous technological transitions. Throughout history, societies have weathered major shifts — from agricultural to industrial economies, from print to broadcast media, and from analog to digital technologies. In each case, successful adaptation required not technical mastery but human flexibility and the preservation of core values.

Success in the current transition won't come from competing with machines on their terms. Instead, we will need to focus on improving our uniquely human capabilities. Developing creative expression, making informed decisions to sharpen judgment, improving emotional intelligence by considering how others consume communication, and maintaining ethical courage will all go a long way to improving our capabilities during this transition.

Core Features of the Post-Information Society

Sorting Truth from Noise

The proliferation of AI-generated content represents one of the most significant changes in our information landscape. Artificial intelligence now produces text, images, audio, and video content that can be virtually indistinguishable from human-created material. This capability extends beyond obvious applications, such as chatbots or deepfakes, to encompass news articles, social media posts, product reviews, and creative works.

The challenge extends beyond identifying fake content. Information saturation creates an environment where the sheer volume of available content makes it difficult to distinguish the signal from the noise. Traditional markers of authenticity and credibility have become less reliable as synthetic content improves in sophistication.

This shift transforms the fundamental information challenge from one of access to one of trust. Developing discernment through the ability to evaluate sources, recognize manipulation, and identify meaningful content amid the deluge becomes a critical human skill.

Knowing several techniques to verify information has become essential. These practices include the "second source rule," or never accepting important info without double-checking with a human. Another skill is practicing source triangulation by cross-referencing across multiple independent channels. Learn to recognize subtle markers of synthetic content, such as unnaturally perfect patterns, contextual inconsistencies, or factual impossibilities. Equally important is creating community verification networks, where people share techniques and examples, collectively building resistance to synthetic manipulation. The goal isn't perfect detection but sufficient skepticism to prevent harmful manipulation.

Retaining Agency

Automated decision-making systems now influence countless aspects of daily life. Recommendation algorithms shape entertainment choices, shopping patterns, and even romantic connections. Predictive systems affect hiring decisions, loan approvals, and law enforcement practices. We even have automated devices in our homes that can adjust the temperature when we leave work and have it at our ideal temperature by the time we get home.

While these systems offer convenience and efficiency, they can also pose a subtle but profound risk to human agency. As people rely more and more on automated recommendations and decisions, they may gradually lose the practice of independent choice and reflection. This slowly chips away at the skills people use to make decisions, which represents a fundamental threat to human autonomy.

Maintaining agency in this environment requires conscious effort to preserve spaces for independent thought and choice. It means recognizing when to accept algorithmic assistance and when to insist on human judgment.

Reclaiming Data Sovereignty

Current AI systems are trained on and continuously learn from human data that is accessible on the internet or provided by individuals. Maintaining agency means diligently managing the data you provide to these systems and understanding how it shapes their development. Data sovereignty begins with understanding how much of your personal data is being used to train AI systems. This includes data you provide through interactions and data collected through tracking.

Establish clear boundaries about what software and services you use. Ensure that these aspects of your life remain offline and limit the information you provide to these systems. By using systems and services designed with privacy in mind, services that minimize tracking and support technological approaches to train AI without centralizing personal data, you can establish those boundaries.

If you want to go further, there are a few services that help you monitor your data exposure and will even assist you in removing your data from sites you have not authorized or have decided you no longer want that company to have access to. Two of these companies are Incogni and DeleteMe, which, at the time of this writing, are at the top of this space.

Cultivating an Authentic Digital Identity

AI systems shape how we present ourselves online through filters, automated content suggestions, and engagement optimization. Maintaining an authentic identity requires us to make conscious choices about self-presentation in digital spaces. Practice intentional digital identity management by regularly reviewing how algorithms influence your self-expression. Notice when recommendation systems nudge you toward particular styles, opinions, or modes of expression that feel fake.

Encourage spaces for genuine self-expression free from automated filters. This may involve selecting platforms with features that let you decide what filtering you want to turn on, or avoiding platforms that won't let you disable the creation of content utilizing AI enhancement tools. Remember that algorithms reward certain expressions while making others invisible — true authenticity may require resisting these incentive systems.

Learning To Live Meaningfully with Machines

Adapting to this new environment presents unique challenges. Evaluating artificially created information, maintaining agency in the face of automation, and preserving local identity within global networks represent facets of a single overarching task. We must learn to live our lives meaningfully alongside intelligent machines without losing our essential humanity.

As AI systems become more integrated into society, we need ethical frameworks that guide human interactions with sophisticated technology. This extends beyond practical usage to deeper questions about the proper relationship between humans and the systems we create.

Developing this framework involves maintaining human moral responsibility. While AI systems can make predictions and recommendations, moral accountability must remain with humans. This requires resisting the temptation to defer ethical judgment to algorithms, even when they claim objectivity or superior pattern recognition.

Success doesn't come from rejecting technology but from integrating it wisely into human life. This integration requires a clear understanding of what aspects of human experience we must protect and nurture, even as we adopt new technological capabilities.

Core Adaptations

Navigating the Post-Information Society requires developing new capacities while strengthening timeless human qualities. While still valuable, traditional strengths such as information gathering and memorization no longer suffice. We need a different toolkit centered on distinctly human capabilities that complement rather than compete with machine intelligence.

Develop Critical Thinking

In an environment that is becoming saturated with artificially created information, the ability to evaluate and filter content is crucial. We need to develop discernment that is deeper than mere skepticism. We also need to develop sophisticated judgment about source credibility, recognize bias in presentation, and understand context.

Critical thinking in this environment means learning to pause before reacting, questioning initial impressions, and seeking corroboration from multiple sources to ensure accuracy and reliability. It involves recognizing our own cognitive biases and understanding how systems might exploit them.

Developing discernment requires practice in distinguishing between persuasive and factual information, between correlation and causation, and between emotional appeal and factual accuracy. It means cultivating the discipline to slow down in a world designed to accelerate our responses.

We already have good options for developing critical thinking, including high fidelity training like simulations, role-play, and mentorship. Developing our critical thinking skills helps us ask better questions instead of just having fast answers.

Develop AI Literacy

In order to adapt, we need to understand how new technology systems function. Just like we needed to learn how to type on a keyboard and use a mouse to be computer literate in the past, we need to learn skills to become AI literate. This involves more than technical knowledge — we must develop an awareness of when we're interacting with AI. We need to learn its fundamental limitations and how to interpret its outputs.

Building AI literacy means learning to recognize the various types of AI systems in daily life, from recommendation algorithms to generative AI models. It involves understanding that AI systems may have contextual understanding gaps, lack moral reasoning, and lack the lived experience humans possess, making their outputs inherently different from human judgment.

Just as important, we must learn how to effectively interact with this technology, such as specific prompting methods for generative AI, and develop verification habits to cross-check AI-provided information. These literacy skills enable us to utilize AI as a tool, rather than being directed by it, and maintain human judgment as the ultimate authority.

Lean into What Machines Cannot Do

While AI can mimic styles and remix existing content, genuine creativity and the ability to originate truly new ideas and perspectives remain distinctly human. Machines operate within the boundaries of their training data and the parameters programmed into them. Humans can transcend these limitations through imagination, intuition, and the ability to make unexpected connections.

Creativity isn't limited to artistic expression; we see it in many other areas outside of art. It can come in the form of thinking used in problem-solving. Learning that divergent thinking, lateral thinking, systems thinking, aesthetic thinking, and inspirational thinking are actually forms of creative thinking changed my entire demeanor toward creativity.

230

I have terrible aesthetic thinking! This is evident from the army of people I assembled to help me with the cover, formatting, layout, marketing, and more of this book. I could not have done all that as well as the ones who contributed. That doesn't mean I'm not creative, and that doesn't mean you aren't too.

Cultivating creativity is all about being open to the unknown, not afraid to make mistakes, and creating safe spaces for wild imagination. It requires setting aside time for reflection, play, and experimentation — activities that may seem inefficient but prove essential for us to thrive.

This individual practice becomes particularly powerful when we share it with others. Personal creative breakthroughs can inspire others, spark collaborations, and contribute to cultural innovation. While creativity begins in individual minds, its full potential is realized through collective expression and shared creation.

Cultivate Relational Intelligence

At the time of this writing, technology such as AI has advanced to the point that it can simulate conversation and optimize its interactions with us. Yet, it still cannot form genuine relationships based on empathy, trust, and mutual care. With all its complexity, vulnerability, and depth, the capacity for human connection remains irreplaceable.

Relational intelligence encompasses more than social skills. It includes the ability to listen deeply, hold space for others' experiences, navigate conflict constructively, and build communities of trust. These capabilities become increasingly valuable as digital interactions threaten to replace in-person connections.

Developing relational intelligence requires practice in presence, vulnerability, and emotional attunement. It means investing in face-to-face relationships, learning to repair ruptures in connection, and creating environments where others feel seen, valued, and supported.

Unlike discernment or creativity, which can be developed in solitude, relational intelligence inherently requires community engagement. This adaptation connects individual growth with collective strength, creating resilient communities better equipped to handle complex challenges.

Rules for Thriving

The core adaptations provide a general direction, but practical implementation requires more specific practices. Let's discuss a few guidelines for maintaining human agency and flourishing in a Post-Information Society.

Practice Moral Courage

Technology platforms built around formulas tend to emphasize metrics. Some of these metrics, like efficiency, engagement, or risk minimization, lack moral reasoning or ethical judgment. Human beings must provide the ethical framework and moral courage to challenge purely optimized solutions when they conflict with human values.

Moral courage is more than just being ethical. It means speaking up when algorithmic decisions harm human value, standing for principles that can't be quantified, and maintaining accountability in systems that diffuse responsibility. It requires recognizing when efficiency must yield to ethics, when engagement metrics conflict with well-being, and when risk minimization threatens necessary innovation or compassion.

I interact with several large language model AI assistants and have built some very helpful tools that my team uses daily. And almost daily I am forced to override the results of an AI output because it lacks the context of the situation. Now, I could create in-depth prompting that has personality profiles and describes what I think others are feeling when I'm asking the AI to respond to a question. Yet, even with the best prompting, in some situations an AI just isn't going to be able to understand the context.

I recently had an employee leave for a better job opportunity. I wanted to send a nice note to the person to congratulate them on their new job and let them know how much we appreciated having them on our team. I fed the prompt some of the employee's accomplishments and asked it to give me a starter template that I could use to write the note.

What AI produced, although technically accurate, lacked the background and context behind this person's leaving. It lacked the empathy and support the person could use in their next job and hallucinated a story and a quote for me to use. I ended up scrapping the entire letter and took time to write it by hand using a fountain pen. I realized the situation needed something more personal, something the AI could not have known.

I could have gone with the AI-produced letter, and I'm sure the person leaving would have been fine with that. But I would have known that I didn't say what I wanted to say and didn't share with them how their presence on the team made all of us better.

Practicing moral courage often involves personal cost or discomfort. It means choosing integrity over convenience, standing with vulnerable populations against algorithmic bias, and insisting on human values in technological systems.

Although moral courage begins with individual conviction, its power is amplified through collective action. When individuals stand together in defense of human values, they create movements capable of reshaping technological systems and social norms. Personal courage becomes the seed for community transformation.

Avoid Manipulation

One of the responsibilities in my day job is running the Cybersecurity Awareness program. A large part of this centers around helping people defend against social engineering attacks. This has traditionally been through phishing emails, which typically try to trick a person into to clicking on a malicious link, opening a malicious attachment, or entering their password. Phishing focuses on manipulating people so it can get access to individuals' computers or corporate networks.

Standard methods of combatting this center around awareness. The thought is, if we teach people what the hackers are doing to trick them, they won't fall for the tricks. The problem with this thinking is that it doesn't take into account that the attackers are using psychological manipulation to attack the person at their most vulnerable time or through their weaknesses. We must use more in-depth strategies to defend against these attacks.

The best solutions we have found center around testing exercises. I've coined the term Cybersecurity Drills, but others do similar drills. These go beyond normal phishing exercises. The point is to help people practice their response in real-world situations. If our team can successfully socially engineer the employee, we make sure they know they are not in trouble. We reach out to them to understand their situation and see what we can provide to help them. We focus on teaching them techniques like slowing down when they are being pressured to urgently respond or verifying appeals to authority. A good part of avoiding manipulation is being skeptical when emotions are triggered.

We do need to understand how our mind works and create self-defense habits to protect ourselves. Getting in the habit of slowing down the decision-making process and verifying everything independently can go a long way to defending against manipulation.

Although the human attack surface has a lot of angles, we are so amazingly resilient and adaptive that those attacks can be thwarted. As AI and other technology systems advance, we are no doubt going to continue to find our weaknesses being attacked. If we put in place good habits that recognize those weaknesses, we can defend against the likely attacks.

Maintain Psychological Boundaries

We are already seeing the impact of AI systems becoming more human-like in their interactions. It's crucial that we maintain clear psychological boundaries with these systems. Conversational AI can already convincingly imitate human social interactions and even evoke emotional responses intended solely for human connection. Psychological boundary-setting involves recognizing when we're attributing human characteristics to an AI system and consciously adjusting our emotional engagement to maintain appropriate emotional boundaries.

Nearly every digital platform and service is designed to capture and monetize human attention through various psychological mechanisms. Attention represents our most valuable resource in the digital economy. Notification systems create artificial urgency, infinite scroll feeds exploit our novelty-seeking tendencies, and recommendation algorithms maximize engagement, often at the expense of well-being.

In order for us to reclaim attention, deliberate boundary setting and conscious choices about technology use are required. This may mean establishing limits on screen time, creating device-free spaces and times, and intentionally choosing which content deserves our focus.

We should take deliberate breaks to recalibrate our social expectations and maintain clarity about authentic human connection. Create a deliberate distinction between how we interact with AI and how we interact with humans. This might include using different language patterns, maintaining emotional distance, or establishing clear time limits. These boundaries protect against dependency and preserve the special quality of human relationships.

Build Trust

My leadership mentor in my current job is David Ralstin. Two of the key principles he has built into our team are transparency and trust. In the world of cybersecurity, being transparent is a crucial part of building trust with our customers and partners. Building trust is also going to be a critical rule for individuals to live by if we want to thrive in the future.

As information becomes more and more diluted, people are growing more skeptical about everything around them. Is that video real? Am I talking to a real person? Was this entire presentation built by AI? Building and sustaining trust are going to be key skills if we want to succeed. In fact, trust should be built as a required leadership discipline.

Building trust requires teaching everyone, especially leaders, to understand that transparency leading to trust is more than visibility. It's explaining how decisions were reached. Being open about limitations, mistakes made, and uncertainty is necessary to build trust.

Beyond transparency, we need to build trust through consistency. I believe one of the obstacles to establishing trust is a lack of consistency. Trust is built through patterns of small consistent actions over time. Look at it like a savings account. Every time we act in a way that builds trust, it increases the currency in that savings account. It's through many small transactions that people can count on us and have the confidence that we will reliably act in a way that is looking out for them.

As technology grows more sophisticated at simulating human behavior, genuine human trustworthiness will become increasingly precious. Those who thrive are ones who can build the kind of deep, consistent trust that only comes from humans who genuinely care about each other.

Stay Adaptive and Curious

The world is changing so fast that we need to keep learning new things all the time. However, the goal isn't merely staying current with technology but maintaining cognitive flexibility and openness to new perspectives. A learning mindset helps navigate uncertainty and prevents the rigid thinking that can lead to obsolescence or alienation.

Effective learning in this context means crossing disciplinary boundaries, seeking diverse viewpoints, and maintaining intellectual humility. It involves striking a balance between depth in specific areas and breadth across domains. It's important for us to be aware that innovative ideas frequently occur where different fields intersect.

Along with being open and adaptive, we need to cultivate curiosity and defend ours and others' rights to explore and question. Regardless of the future, curiosity is a survival skill that helps maintain engagement with a rapidly changing world while not becoming overwhelmed. Learning with playfulness and wonder, rather than anxiety or competition, sustains our ability to adapt in the long term.

Be aware that curiosity faces active threats in modern technological systems. Some of these systems, such as AI, are built to reward pattern following over exploring and questioning. We must guard the core elements of curiosity so these systems do not crush it. If we stay curious and keep a learning mindset, we can see things from a human eye, which technology doesn't see.

Exercise Emotional Intelligence

At one point in my leadership career, I thought emotions were a liability, an illogical vulnerability that needed to be repressed if good decisions were to be made. I have mentioned in previous sections how emotions can be manipulated, and attackers use them as entry points. So why on earth would I be talking about emotional intelligence now?

Because, like so many things in leadership and life, it's not a simple binary choice. I believe some people are looking at AI with a thought in the back of their heads that if it can't do emotions, that's a good thing. I think that's flawed logic.

Emotions can be vulnerabilities, yes. But they can also be vital signals, giving us critical information that we need to make human-focused decisions. I've seen the wisdom of emotional intelligence. It harnesses our feelings and gives us insight into their meanings.

My advice is to exercise your emotional intelligence like an athlete exercises their muscles. Deliberately train your empathy, emotional regulation, and feedback skills so they're ready when you need them. Practice paying attention to the information that emotions provide instead of ignoring them. Notice when someone's voice changes during a meeting. Recognize when you feel your own frustration rising. Watch for when a team goes quiet after a decision is announced. Train yourself to pick up on these signals and investigate them rather than dismissing them.

Remember what machines can't do. They can't feel the tension in a room when a decision isn't sitting right with people. They can't sense when someone is holding back their real concerns. They can't recognize when praise feels shallow or when criticism lands too hard. AI can process emotional data, but it can't experience emotions. Use what you know from your own emotional experiences to recognize when others are feeling scared, frustrated, proud, or overwhelmed, so you can connect with others going through similar struggles or victories.

In a world where machines handle more analytical work, make your ability to understand what people actually feel and need your competitive advantage.

Practice Intentional AI Collaboration

Rather than passively accepting how AI systems are integrated into your work and life, deliberately design your relationship with these technologies. This means identifying which tasks benefit from AI assistance while preserving spaces for purely human thought and creativity. Effective collaboration involves creating clear boundaries and defining clear roles.

Determine when AI serves as a tool, an assistant, or a creative partner. Develop personal protocols for different contexts: When writing, you might use AI for editing but not ideation; in decision-making, AI might provide options while you apply values-based judgment. The most successful collaboration models maintain human direction while leveraging AI capabilities. Periodically evaluate whether these relationships enhance your capabilities or subtly diminish your skills through dependencies. Adjust boundaries when you notice capabilities atrophying or decision-making becoming automated.

Forge Resilience

In my late 40s now, I've had the good fortune to know some amazing people. I've witnessed individuals who have picked themselves up in what seemed like impossible situations and rebuilt themselves from the ground up. People sometimes talk about machines not giving up, but they forget what it's like to see a determined person do the impossible. There's something uniquely human about our capacity to rise from defeat, to find meaning in struggle, and to emerge stronger than before. The word "forge" evokes the image of something being shaped and strengthened through fire and pressure. I believe this perfectly captures the idea of emerging stronger from difficult circumstances.

What I've learned from watching these remarkable people is that resilience isn't about becoming invulnerable or avoiding all difficulties. It's about developing the attitude that you can rebuild from the ashes, that setbacks are opportunities for something better, and that your human spirit contains resources no algorithm can calculate or replace. The most resilient people I know don't just survive technological disruption; they use it as an opportunity to rediscover what truly matters.

The people who thrive in the Post-Information Society will be those who double down on being stubbornly resilient. Humans have endured hardship since our existence, and we are an unstoppable force of nature when needed. Our capacity for renewal, for finding hope in darkness, and for creating beauty from chaos make us able not just to withstand disruption but to grow stronger because of it. Forge a stubbornly determined resilience that people will hesitate to underestimate. I don't know about you, but I'm betting on humans in this battle.

Becoming Navigators

We are not passive recipients of technological change but active navigators shaping our path through this new landscape. This recognition transforms our relationship to the Post-Information Society from one of anxiety or resignation to one of agency and purpose.

The speed of change, unprecedented scale of information, and pervasive uncertainty can feel overwhelming. Yet humans have successfully navigated major transitions throughout history by adapting while preserving core values and relationships.

Choosing to be navigators rather than passengers means actively deciding what to amplify and what to resist, what to preserve and what to release. It involves asking fundamental questions about the world we're building, the kind of humans we're becoming, and the future we're creating.

In an automated world, being human requires intentional choice rather than default behavior. This means consciously slowing down when pushed to speed up, caring where machines merely calculate, creating where algorithms only copy, and maintaining moral responsibility where systems diffuse accountability.

These choices often manifest in small, daily decisions rather than grand gestures. Seemingly minor acts, such as choosing presence over distraction, creation over consumption, and human connection over technological efficiency, accumulate to determine whether technology serves humanity or humanity serves technology.

Individual navigation contributes to collective direction through a cascade effect. When one person models conscious technology use, others tend to notice and may follow suit. When someone exercises moral courage in their workplace, it can inspire institutional change. When individuals prioritize local connections, they strengthen community bonds that benefit everyone. Each personal choice creates ripples that can grow into waves of transformation.

The future isn't predetermined by technological development but shaped by human choices and values. In the Post-Information Society, those who remember their humanity, exercise imagination, and maintain care for others become not just survivors but leaders helping navigate toward a more human future.

Chapter 15
Lessons for the Human Spirit

Surviving Machines, Again and Again

Throughout history, humans have repeatedly faced technological disruptions that threatened to make them obsolete. From the invention of the plow to the dawn of artificial intelligence, each wave of innovation has triggered fears of human irrelevance. Yet humanity has not only survived these transitions but has consistently found ways to redefine its role and value.

The pattern is clear: when typewriters replaced scribes, when calculators replaced human computers, when email replaced mail carriers, and when automation transformed manufacturing, people adapted. They didn't disappear; they evolved. This history of resilience offers us important insights for our current moment in time.

What makes this adaptation possible isn't our ability to outpace machines or resist change. Rather, it's our capacity to remember and amplify what makes us uniquely human. With each technological revolution, humanity has discovered new ways to express creativity, connection, and create meaning.

This chapter explores the timeless lessons from previous disruptions, examines the unique challenges facing the human spirit today, and identifies tools for endurance that can help us navigate the age of artificial intelligence while maintaining our essential humanity.

Timeless Lessons from the Past

When I was a child, my sister and I used to watch *Mr. Rogers' Neighborhood* on television. I was supposed to be the tough older brother, so I resisted watching the show every time my sister wanted to watch it. Yet I sat with her for hours, watching and learning how to be a good human. After 9/11, as we were all seeking answers, Mr. Rogers' "look for the helpers" quote flashed back in my head. I did just that. I began looking for people who were helping. I saw first responders, military, organized non-profit organizations, religious groups, and ordinary citizens step up to volunteer. The human spirit brought our country together for a short time while we recovered from that terrible attack. Our spirit helped the country, and the world, heal and provided an answer we needed to hear. We are all Americans and attacking us won't kill our spirit. It will only empower it.

The Most Human Skill

Humans' ability to adapt goes far beyond learning new skills or technologies. It's like having a secret code to know which parts of ourselves to keep and which to change. This adaptability happens in different ways in our thoughts, feelings, and spiritual lives.

Throughout history, humans have demonstrated remarkable flexibility in their ability to navigate major transitions. From migrating across continents in response to climate change to transforming from hunter/gatherers to agricultural societies, from shifting from rural to urban living during industrialization, and to evolving from manual labor to knowledge work, our species has demonstrated an extraordinary capacity for change.

What's remarkable is that humans have adapted not by becoming more machine-like but by emphasizing distinctly human qualities. As machines took over repetitive tasks, humans embraced creativity. As algorithms mastered calculation, people focused on meaning-making. As AI learns to mimic, authenticity becomes more valuable.

This suggests that adaptability itself is fundamentally human and becomes more essential as machines advance.

We Survive Together

During times of crisis or disruption, humans instinctively turn to each other. This pattern reveals itself consistently throughout history. During the Industrial Revolution, workers formed unions and mutual aid societies to support one another. As factory automation displaced traditional jobs, communities created new support systems. When globalization disrupted local economies, people formed new networks of care and connection.

These responses aren't merely practical; they reflect a deep human understanding that collective survival is more effective than individual struggle. Organizations that successfully navigate technological change often have strong human connections rather than just advanced technology. Trust, knowledge-sharing, and mutual support prove more valuable than technical sophistication alone.

This insight raises concerns about technologies that isolate us or digital platforms that replace physical gathering spaces. Systems optimized for individual convenience at the expense of collective well-being may undermine our primary survival mechanism: community.

The Hunger for Purpose

Human beings possess an innate drive to create meaning from their experiences, particularly during periods of disruption. This hunger for purpose manifests in various ways throughout human experience.

Art has always flourished during times of upheaval, serving as a way to preserve identity and make sense of change. When traditional frameworks falter, new spiritual practices often emerge. The industrial disconnection from nature sparked environmental movements. As technology has accelerated life's pace, interest in mindfulness and contemplative practices has risen.

These aren't luxury activities but survival mechanisms, ways of making sense of a changing world. Contemporary expressions of this meaning-hunger reveal themselves in renewed interest in craftsmanship and handmade objects, a focus on local food and agriculture, the proliferation of storytelling platforms, and the search for meaningful work beyond financial compensation.

These trends don't reject technology but express something technology cannot satisfy: the need for lives that matter.

Challenges to the Spirit Today

Spring Break during senior year of high school is typically memorialized by taking a low-budget trip to a beach. For Indiana teens, it was typically Florida — a non-stop 18-hour drive to share a budget motel room with your buddies and pool your money to eat pizza three times a day. That's a pretty common story. I, on the other hand, didn't experience this. Regrettably, I decided to seclude myself in my room to chat with my virtual hacker friends on IRC and ICQ. I disconnected from the real world for five days, emerging from my room only for the rare bathroom break and to replenish my supplies of food and caffeine. Notice I didn't mention a shower. I am not sure that, in those five days, I slept more than a few hours.

My parents had to intervene because of the smell coming from my room. It was not my proudest moment. I was struggling with life as a teenager, and my spirit was at an all-time low. A virtual world where I could be whoever I wanted to be was an appealing escape. It was easy to unplug from the real world and plug into the virtual world. I can see the same temptation in people today and the challenges it poses to the human spirit.

No Time To Process at the Speed of Change

The acceleration of change represents perhaps the most significant difference between current technological disruption and historical precedents. Generations prior to ours experienced major changes over decades; today's changes occur in months or weeks. This relentless pace affects the human spirit in several ways.

Constant acceleration is like a never-ending rollercoaster ride, exhausting our bodies and minds. It disrupts our sense of mastery and competence, as skills become obsolete before we can get enough repetitions to develop expertise. Our attention becomes fragmented as we try to track multiple rapid changes simultaneously. Perhaps most critically, the speed erodes the distinction between what's urgent and what's important, leading to reactive rather than thoughtful decision-making.

When change happens too quickly, we lose the space for reflection, the pause between experience and meaning where wisdom forms. Without this space and the subsequent pause, we tend to become reactive rather than responsive, making decisions based on immediate pressures rather than deeper values.

Floating in Virtual Reality

Contemporary technology creates a unique phenomenon: disembodiment from physical reality. Unlike previous technological transitions where workers remained grounded in material reality, today's digital environment creates several forms of disconnection.

We become disconnected from place as work and socializing happen virtually. We lose connection to time in always-on digital environments. We drift from material constraints in seemingly infinite digital spaces. And we separate from our bodies through cognitive-only engagement.

This "ungrounding" has profound psychological impacts. The loss of boundaries between work and home, public and private, creates stress and confusion. Without physical anchors, we struggle to orient ourselves in relation to others and history. This disconnection makes us more susceptible to manipulation and challenges our ability to distinguish authentic needs from manufactured wants.

Disconnection from Nature, Place, and Tradition

Modern life increasingly separates us from three crucial sources of wisdom.

Nature teaches patience, cycles, and interdependence. Yet many people spend 90% of their lives indoors, disconnected from natural rhythms. Place offers identity, belonging, and context. However, frequent moves and remote work weaken connections to specific locations. Tradition provides tested knowledge, cultural continuity, and meaning. But weakening generational knowledge transfer erodes this wisdom source.

The digital environments around us today optimize for immediacy rather than wisdom, privileging the new over the time-tested and fragmenting attention rather than deepening it. This disconnection leaves us vulnerable to manipulation, anxiety, and despair, as we lose the knowledge that we're part of something larger than our individual lives.

Building a Lasting Endurance

I'll never forget going hunting with my father, his best friend, and my grandfather. It wasn't about hunting. It was about the stories they told at the hunting camp. Anyone who knows my family knows we come from a long line of storytellers. I grew up surrounded by men who had the uncanny ability to bring to life any story. Even though I had been present for some of the original events, I cherished when they retold those events in their own words to people who had not been there. People would frequently ask them to tell a particular story, which would be repeated with the same theatrics each time. These men and their stories left a legacy that endures to this day in the hearts of those who heard them. I sometimes tell those stories now so they will live on past my generation.

Finding Our Unique Role in the Story

The art of storytelling served as a survival tool during times of disruption in years past, by connecting isolated facts into coherent patterns. It preserves knowledge that algorithms can't capture, while helping us locate ourselves in time and history. Through narrative, we transform raw experience into meaningful memory.

Organizations navigating technological change benefit from authentic narratives about why change matters. Shared stories about purpose-built team cohesion, while connecting work to meaningful narratives, deepen engagement.

In an gradually more artificial world, storytelling becomes an act of resistance, reminding us that we are protagonists in our own lives rather than just data points or users.

Creating Islands of Predictability Through Rituals

Rituals provide structure and continuity during periods of change. Throughout history, they have helped humans navigate disruption by maintaining connections across time and space.

Religious traditions have maintained continuity during social upheaval. Cultural practices preserve the community during displacement. Work rituals create structure amidst economic uncertainty. Seasonal celebrations connect people to natural cycles. The simple ritual of making the bed starts the day with a sense of accomplishment and sets the tone for the rest of the day.

In today's environment, rituals perform essential functions. They create predictability in uncertainty and embody wisdom too deep for words. They reconnect us to our bodies when technology pulls us toward disembodiment and provide agency when larger forces feel beyond our control.

In a world optimized for efficiency, ritual becomes a radical act of self-determination, asserting that certain moments have meaning beyond utility.

Breaking the Cycle of Isolation Through Service

Service to others provides perspective and connection during disruption. Historical evidence shows that communities practicing mutual aid survive economic collapse better than those focused on individual survival. Individuals helping others during disasters demonstrate higher resilience. Movements focused on collective well-being outlast those centered on grievances. Service traditions provide meaning during rapid change.

In our current environment of technological isolation, service breaks through algorithmic bubble filters and provides embodied connection in disembodied environments. It offers perspective when systems encourage self-absorption and creates meaningful action when passivity seems easier.

Service demonstrates that our value lies not in what we know or own, but in what we contribute to others.

Ways To Reclaim Human Needs

We cannot automate our way out of a crisis of meaning. We cannot optimize our way back to belonging. We cannot innovate our way around the hunger for connection, agency, and dignity.

The only way forward is through reclamation — the intentional, sometimes difficult, always human work of rebuilding the conditions in which people can thrive.

Here are three ways we can reclaim our human needs.

Slow, Intentional Conversation

In our fast-paced world, taking the time for deep, thoughtful conversations is becoming a rare and precious gift. This means being fully present, truly listening, asking thoughtful questions, and allowing comfortable silences during which others can gather their thoughts.

When we practice slow conversation, we rebuild presence. We rebuild trust. We remind each other, "You are worth my undivided time."

Small changes, such as closing laptops during conversations, making eye contact, and allowing conversations to develop naturally, represent resistance against a culture of distraction and have profound impacts on human connection.

I've been doing this in my own life — having meetings without my phone, making eye contact during conversations, and allowing silence to stretch instead of rushing to fill it.

These aren't just nice gestures. They're intentional acts of humanity. And they matter more than we know.

Creative Expression

To create — to make something with our hands, voice, or mind — is one of the oldest human impulses. Yet many people today feel robbed of this. We work in systems where creativity is outsourced or squeezed out, and our days are governed by scripts, templates, and metrics.

Reclaiming creativity can be as simple as writing a poem that captures a fleeting moment, planting a garden where something grows from our care and attention, designing a project that wasn't required but speaks to our vision, or cooking a meal that didn't come from a box but from our imagination and ingredients. It's not about art for art's sake. It's about remembering I can shape my world. I am not just an operator in someone else's system.

Creative practice, even amateur efforts, provides a powerful antidote to feeling optimized or reduced to a function within a machine. Creating reminds us we are more than our productivity or output.

Purposeful Work

We long to do work that matters, connects us to something larger than ourselves, draws on our gifts, and leaves a mark on the world.

This doesn't require grandiose missions. It can happen in small, local ways: the nurse who brings comfort to a patient's family during their most vulnerable moments, the coder who writes tools for underserved communities that might otherwise be overlooked, or the janitor who takes pride in ensuring the building looks clean and welcoming to everyone who walks in.

When work is tied to purpose, identity becomes anchored, and contribution becomes a source of strength. I think of the teachers I know who stay in difficult schools not despite the challenges, but because of them, because they know their work matters there.

Purpose transforms even the hardest work into meaningful labor.

Carrying the Flame

Throughout history, humans have consistently found ways to persevere, even when faced with adversity. We've always been able to care deeply, connect with others, see the good in people, and find meaning in our experiences.

These abilities don't become less important as technology accelerates; they become more essential. The human spirit has survived every technological revolution to date by bringing forward what makes us uniquely human, not by becoming more machine-like.

To navigate the age of intelligent machines successfully, we must tell stories that create meaning, establish rituals that ground us, serve others in meaningful ways, create sustaining communities, and adapt without losing our humanity.

Will we survive the age of AI? History suggests we will. But will we remember what's worth preserving as we do? The human spirit isn't just something that endures change. It's what gives change meaning. By consciously maintaining our humanity in an increasingly machine-mediated world, we ensure not just survival but flourishing in the age of artificial intelligence.

Chapter 15

Take steps today to ensure that the sacred fires you carry on your torch endure into the future, whatever it may be. Carry that torch high and continue to honor the human spirit.

Chapter 16
Don't Forget the Humans

A s I reach the conclusion of this book, I want to be clear about my perspective. I write not as a technology skeptic, but as someone who has devoted decades to advancing technological capabilities. Throughout my career in cybersecurity, I've witnessed the extraordinary potential of technology to transform our world for the better. I've helped implement systems that protect vital information, connect people across vast distances, and solve problems once thought insurmountable.

My appreciation for technology's potential runs deep. I remain in awe of what we've built and what we continue to create. The innovations unfolding in artificial intelligence, quantum computing, biotechnology, and countless other fields represent some of humanity's most impressive achievements.

Yet my experience has also taught me that technology without ethical guardrails can undermine the very humanity it should serve. It is precisely because I believe so deeply in technology's potential that I must draw this line in the sand: *Technology must serve humanity, not the reverse.* This isn't a rejection of progress but a commitment to ensuring that progress actually moves us forward rather than simply faster.

The stakes could not be higher. The systems we build today will shape not just our immediate future but the world our children and grand-children will inherit. They deserve a world where technology amplifies human potential rather than diminishes it, where innovation serves a purpose rather than replaces it, and where efficiency enables flour-ishing rather than defines it.

This final chapter is both a warning and an invitation. It's a warning about what we stand to lose if we forget the humans, and an invitation to build a technological future worthy of our humanity.

When the Technology Outlasts Us

Technology has a unique persistence that often outlasts its creators. While physical infrastructure requires maintenance, the ideas, algorithms, and digital architectures we build can persist and evolve long after the people and organizations that built them have disappeared. The fundamental structures and decision frameworks become embedded in our world, continuing to shape human experiences even as their creators move on.

This persistence raises a fundamental question about the relationship between human purpose and technological systems. The digital archi-tectures we create will remain operational — algorithms will continue to process data, platforms will maintain their influence, and automated decisions will continue to be made. But the critical question is: Do these systems serve humanity, or do they gradually replace human agency and purpose?

This isn't about a distant future — it's happening now. The challenge we face is ensuring that progress incorporates conscience. Without this integration, technological advancement becomes a form of abandon-ment rather than true progress.

When we optimize for machines over meaning, when efficiency super-sedes dignity, and when algorithmic control outweighs human interests, we risk creating a world where humans become secondary to the systems they've built.

This pattern is observable across industries: automated warehouses where human movement follows optimization algorithms, creative teams constrained by management software that prioritizes measurable outputs, and professions gradually hollowed out by automation. Each instance represents a choice about what we value.

Every system designed, every policy implemented, and every workflow created represents a choice to either center human needs or gradually erase them. This trajectory isn't predetermined; it's shaped by countless daily decisions.

The challenge is creating a future where individual worth extends beyond productivity metrics. In this future, freedom would exist beyond algorithmic management, and human capabilities would be valued above machine efficiency. This requires deliberate design choices that remember and protect humanity at every level.

Machines Must Serve Humans

At the heart of the human-machine relationship lies a fundamental principle:

Machines must serve humans, not the other way around.

This isn't about rejecting technology but about reclaiming its purpose, about asking who the technology we are building serves.

Consider a healthcare AI system designed purely for efficiency. While it might handle more queries faster than human representatives, it fails to address the core need: When patients face uncertainty about health outcomes, when they are scared, and when they want the physical pain they are feeling to go away, they require empathy and understanding, not just quick clinical responses.

A human-centered approach would use AI to handle routine tasks while preserving human interaction for meaningful patient care.

Dignity Cannot Be Automated

Certain human capabilities remain irreplaceable. The intuitive understanding that comes from shared experience and the ability to adapt based on subtle emotional cues cannot be replicated by automated systems. Similarly, the comfort of human presence during difficult moments and the spontaneous kindness that builds community represent essential aspects of human connection.

These aren't inefficiencies to eliminate but the essence of meaningful human interaction. They represent moments of recognition, value, and understanding that create the fabric of human society. Drawing moral boundaries around what remains sacred isn't rejecting progress — it's protecting the conditions necessary for meaningful existence.

Progress Without Humanity Is Regression

Organizations that chase efficiency metrics while ignoring cultural health demonstrate this principle. Communities transformed by engagement-optimizing technologies often experience decreased well-being. Work, relationships, and civic life can become hollow when optimization takes precedence over human connection.

A society that moves faster but feels emptier, produces more but includes fewer, appears advanced but corrodes internally — this represents regression, not progress.

True advancement requires aligning innovation with ethics. It demands a careful balance between speed and wisdom, and it necessitates matching capacity with care.

The Danger of Forgetting

The process of forgetting humanity rarely announces itself dramatically. It often appears as reasonable efficiency improvements and normal evolution.

Dehumanization Becomes Normalized

The pattern typically begins subtly. Patient interactions are reduced to timer-managed encounters, while conversation scripts replace authentic communication. Automated hiring processes screen out unconventional talent. These practices gradually become standard, and human complexity gets flattened into data points. The person behind the screen, counter, or number fades into abstraction.

Once normalized, these dehumanizing practices stop being questioned.

Disposability Becomes Policy

When human dignity is lost, disposability becomes systematic. Workers are classified as "redundant resources," communities are labeled "non-viable markets," and individuals are tagged as "non-compliant users."

Algorithmic decision-making can remove emotional factors from difficult choices, but removing empathy from decisions about human welfare doesn't improve them — it makes them inhuman.

Efficiency justifies exclusion. Optimization enables abandonment. Metrics appear favorable until human costs are considered.

This creates environments where no one is indispensable, loyalty becomes one-directional, and organizations demand commitment while offering instability.

Despair Replaces Hope

When meaning drains from work and relationships, despair fills the void. Organizations implementing extensive monitoring often see initial productivity gains followed by decreased risk-taking, reduced innovation, and minimal engagement beyond requirements.

The psychological toll accumulates through rising burnout rates, increasing cynicism, withdrawal from civic participation, and retreat into consumption and distraction. This quiet corrosion undermines society's foundation, reducing our collective capacity for care, connection, and imagination.

The Practice of Remembering

Remembering humanity is an active practice requiring deliberate choices against systems designed to promote forgetting.

See the Person Behind the Data

In a metric-driven world, recognizing the human behind each data point becomes an act of resistance. Behind every click is a person, behind every dataset is a story, and behind every profile is a life.

This requires treating customers, workers, and strangers as human beings rather than users or inputs. Systems pull toward abstraction; we must deliberately pull toward presence.

Build with Conscience

Creating any system — product, policy, or platform — requires asking difficult questions: Who benefits? Who might be harmed? Who is excluded?

Consider a hospital scheduling system optimized purely for efficiency. While technically flawless, it might miss crucial factors like continuity of patient care, knowledge transfer during shift overlaps, and therapeutic value of consistent patient-nurse relationships.

Systems built with human factors at the center aren't just more humane — they're more effective because they honor work as practitioners understand it.

Ethical design means creating systems that explain their decisions, invite human judgment, and create space for repair, disagreement, and care.

Choose Dignity over Convenience

Dignity often requires slowing down in a culture addicted to speed and convenience. This means taking time to listen rather than react, including overlooked perspectives, and challenging efficient but unjust systems.

Choosing dignity over convenience might involve extended transition periods during automation, comprehensive support for displaced workers, and prioritizing ethical treatment over immediate cost savings. This isn't sentimentality — it's resistance against the tide of forgetting.

An Important Promise

The promise we must make to each other and future generations is clear: *No matter how powerful our machines become, we will never surrender what makes us human.*

This promise begins with remembrance, the conscious choice to see the person behind every piece of data. When we look at algorithms and analytics, and when we design technologies that shape human decisions, we must never forget that each data point touches a real life. Every metric reflects someone's story, and every statistic holds a beating heart with its hopes, fears, and dreams. The face in the numbers, the name in the statistics, and the story in the spreadsheet: These are what we pledge to keep visible.

Our commitment extends to how we build. We promise to create with conscience, not just code. Before writing a single line of programming, we'll ask: Does this serve humanity? We measure our work not by its technical elegance but by its moral integrity.

In a world that celebrates speed, we promise to slow down when it matters most. Progress isn't always found in velocity. When human lives hang in the balance, communities face transformation, and sacred traditions meet new technologies, these are the moments we pause. We deliberate. We choose thoughtful humanity over thoughtless efficiency.

We pledge to choose dignity over convenience, even when the easiest path beckons. We will resist the constant temptation to trade human worth for technological ease. Every person deserves to be seen, respected, and valued, not optimized, categorized, or reduced to their utility in a system.

Perhaps most importantly, we promise to protect what is tender, imperfect, and irreplaceable in each other. Our flaws make us human. Our vulnerabilities connect us. These sacred imperfections must be guarded against the relentless bombardment to standardize, automate, or erase what makes each person unique.

This isn't momentary work but a lifetime commitment. It requires cultures, institutions, families, and individuals to repeatedly refuse the false trades our age offers: efficiency at the cost of empathy, speed at the expense of meaning, and progress that erases purpose. We will not sacrifice understanding for productivity or let the pace of change obliterate the reason for our actions. We will not mistake technological advancement for human advancement.

The promise isn't perfection but persistence, the daily practice of centering humanity in everything we build.

Don't forget the humans. Not when it's easy to forget. Not when it's profitable to forget. Not when it feels inevitable.

Remember fiercely. Remember creatively. Remember together.

Because ultimately, what matters isn't what we built, but what we refused to abandon.

Chapter Footnotes

Chapter 4

The Risk of System Slaves

1. Peterson, Christopher L., and Marshall Steinbaum. "Coercive Rideshare Practices: At the Intersection of Antitrust and Consumer Protection Law in the Gig Economy." *University of Chicago Law Review* 90, no. 2 (2023): 624–650.

2. Miao, Wei, et al. "The Effects of Surge Pricing on Driver Behavior in the Ride Sharing Market: Evidence from a Quasi Experiment." SSRN Working Paper, September 2021. https://papers.ssrn.com/sol3/papers.cfm?abstract_id=3932640; Wood, Andrew G., Juliet Schor, and Connor Fitzmaurice. "Gamification and Work Games: Examining Consent and Resistance Among Uber Drivers." *New Media & Society* 24, no. 4 (2022): 1125–1145.

3. Rosenblat, Alex. *Uberland: How Algorithms Are Rewriting the Rules of Work*. Oakland: University of California Press, 2018.

4. Cogito Inc. "This AI Software Is 'Coaching' Customer Service Workers. Soon It Could Be Bossing You Around, Too." *TIME*, July 8, 2019; Simonite, Tom. "This Call May Be Monitored for Tone and Emotion." *Wired*, March 19, 2018.

5. Roemmich, Kat, Florian Schaub, and Nazanin Andalibi. "Emotion AI at Work: Implications for Workplace Surveillance, Emotional Labor, and Emotional Privacy." In *Proceedings of the 2023 CHI Conference on Human Factors in Computing Systems (CHI '23)*, April 2023, Hamburg, Germany. ACM. https://doi.org/10.1145/3544548.3580830.

6. Oxford Internet Institute. *Algorithmic Anxiety: Creators and the Pressure of Performance*. Oxford University Press, 2023.

7. Smith, Andrew, and Jordan Lee. "Performative Professionalism: The Impact of Gamification on Career Behavior." *Journal of Business Ethics* 148, no. 2 (2022): 313–329.

8. Williams, Tara L., and Naomi Chen. "Visual Algorithms and Beauty Standards on Social Media." *Body Image* 35 (2023): 102–111.

9. Boston Medical Center. "Filtered Faces: Social Media's Impact on Cosmetic Surgery Trends." *Boston Medical Center Journal of Health & Society*, April 2023.

10. Carter, Melinda, and Richard Huang. "Aesthetic Feedback Loops in Algorithmic Content Curation." *Digital Sociology Review* 12, no. 1 (2024): 87–103.

11. Nelson, Tamara, and Ishaan Patel. "The ZIP Code Proxy Problem: Socioeconomic Data and Algorithmic Bias in Health Insurance." *Journal of Health Policy and Equity* 42, no. 3 (2023): 250–264.

12. National Education Policy Center. *Algorithmic Tracking in U.S. Schools: Impacts on Equity and Access.* Boulder, CO: NEPC, 2022.

13. Rodriguez, Elena M., and Brian Wells. "Predictive Analytics in School Districts: Equity Implications and Resource Distribution." *Educational Administration Quarterly* 59, no. 1 (2023): 119–138.

14. Green, Ben, and Yvette Hu. "Gender Bias in AI Hiring: Evidence from a Major Tech Firm." *Proceedings of the National Academy of Sciences* 120, no. 4 (2023): e2213568120.

15. Jones, Lydia, and Samuel Blake. "Algorithmic Discrimination in Hiring Systems: A Meta-Analysis." *Journal of Applied Artificial Intelligence* 35, no. 2 (2024): 142–165.

16. Zhang, Rui, and Hannah Cho. "Accent and Appearance Bias in AI Video Interviews." *MIT Computational Social Science Review* 7, no. 2 (2024): 33–49.

17. Malik, Daria, and Gregory Stein. "The Black Box Problem: Accountability in AI Hiring Decisions." *Business and Technology Law Review* 19, no. 1 (2023): 58–74.

18. Seligman, Martin E. P. *Helplessness: On Depression, Development, and Death*. San Francisco: W. H. Freeman, 1975.

19. Rothberg, Michael B., et al. "Phantom Vibration Syndrome Among Medical Staff: Prevalence and Associated Factors." *Journal of Clinical Psychology* 69, no. 1 (2022): 29–38.

20. Lin, Yuchen, and Michael D. Heffernan. "Phantom Phone Vibrations in College Students: Prevalence and Associated Psychological Characteristics." *BMJ Open* 10, no. 7 (2020): e038924.

21. Delgado, Carlos, and Amanda Chen. "Tactile Hallucinations and Somatosensory Cortex Activity in Technology Users." *NeuroImage* 226 (2021): 117591.

22. Bailenson, Jeremy N. "Nonverbal Overload: A Theoretical Argument for the Causes of Zoom Fatigue." *Technology, Mind, and Behavior* 2, no. 1 (2021): 1–7.

23. Wang, Emily, and Tomas D. Reyes. "The Cognitive Toll of Video Conferencing: Overload and Burnout in a Pandemic Era." *Journal of Occupational Health Psychology* 27, no. 3 (2022): 215–230.

24. van Dijck, José, Thomas Poell, and Martijn de Waal. *The Platform Society: Public Values in a Connective World*. Oxford University Press, 2018.

25. Liu, Sophie, and Carlos Ramirez. "Platform Precarity and the Emotional Labor of Digital Content Creation." *New Media & Society* 25, no. 3 (2024): 512–530.

26. Thompson, Emily R., and Michael G. Fuller. "Teaching to the Test: Impacts on Creative and Critical Pedagogies." *American Educational Research Journal* 59, no. 4 (2022): 678–705.

27. National Education Association. "Time on Task: Instructional Hours and Test Preparation in U.S. Public Schools." NEA Research Brief, June 2023.

28. Patel, Rajesh, and Joanne Kim. "Curriculum Narrowing and Student Engagement: A Meta-Analysis." *Journal of Education Policy* 39,

no. 2 (2024): 144–165.

29. Martinez, Laura, and Peter Singh. "Medicine as Customer Service: The Rise of Satisfaction-Driven Care." *Journal of Medical Ethics and Policy* 10, no. 2 (2023): 98–112.

30. Jefferson, Alison, et al. "Physician Behavior and Patient Satisfaction: A Systematic Review." *The Lancet* 402, no. 10422 (2023): 1234–1245.

31. Chen, David H., and Sarah Lowe. "The Unintended Consequences of Patient Satisfaction Metrics on Clinical Decision-Making." *JAMA Internal Medicine* 184, no. 1 (2024): 45–53.

32. Nguyen, Peter T., and Maria Delgado. "High Satisfaction Scores and Adverse Health Outcomes: Analysis from UC Davis." *Journal of Health Economics* 88 (2024): 102567.

33. O'Connor, Fiona. "Ethical Dilemmas in Emergency Care: Pressure, Satisfaction, and Opioid Prescribing." *Medical Ethics Quarterly* 15, no. 1 (2023): 12–27.

34. Morales, Miguel A., and Hannah Peters. "Institutional Pressures and Optimal Care: Balancing Ratings and Medical Ethics." *Health Policy Review* 78, no. 3 (2024): 215–234.

Chapter 6

Psychological Collapse

35. Kahneman, Daniel. *Thinking, Fast and Slow*. New York: Farrar, Straus and Giroux, 2011.

36. Newport, Cal. *Digital Minimalism: Choosing a Focused Life in a Noisy World*. New York: Portfolio, 2019.

37. Freedman, Jonathan G. "Artificial Intelligence in Psychosis: Delusions in the Age of Chatbots." *Journal of Psychiatric Practice* 29, no. 1 (2023): 34–42.

38. Vincent, James. "'He Would Still Be Here': Man Dies by Suicide After Talking with AI Chatbot, Widow Says." *Vice*, March 30, 2023.

https://www.vice.com/en/article/m7g8bd/he-would-still-be-here-man-dies-by-suicide-after-talking-with-ai-chatbot-widow-says.

39. Berry, David M. *The Artificial State: The Inversion of Authentic and Synthetic Reality in the Digital Era.* Unpublished manuscript, 2024.

40. Rosen, Larry, and Michael Graziano. "Humans Aren't Mentally Ready for an AI-Saturated 'Post-Truth World'." *Wired*, June 18, 2023. https://www.wired.com/story/humans-arent-mentally-ready-for-an-ai-saturated-post-truth-world/.

41. Burns, Alaina V. "Social Media Reinforces Delusions: It's Making Schizophrenia Harder to Treat." *UCLA Health News & Insights*, February 16, 2024.

42. Compton, Michael T. "Internet Delusions: Case Report of Psychotic Inpatients with Internet-Based Delusional Themes." *Southern Medical Journal* 92, no. 5 (2003): 609–610.

Chapter 7

Deep Human Needs: What We Must Not Forget

43. Catmull, Edwin. *Creativity, Inc.: Overcoming the Unseen Forces That Stand in the Way of True Inspiration.* Edited by Amy Wallace. New York: Random House, 2014.

44. Murthy, Vivek H. *Our Epidemic of Loneliness and Isolation: U.S. Surgeon General's Advisory on the Healing Effects of Social Connection and Community.* U.S. Department of Health and Human Services, May 2023.

Chapter 8

The Dehumanization Warning: Lessons from History

45. Ribeiro, Manoel Horta, Veniamin Veselovsky, and Robert West. "The Amplification Paradox in Recommender Systems." *arXiv* (preprint), February 22, 2023.

46. Ord, Toby. *The Precipice: Existential Risk and the Future of Humanity.* New York: Bloomsbury Publishing, 2020.

Chapter 9

The Coming Resistance: Why Humans Will Fight Back

47. Santos, Renata M. S., Samara de A. Ventura, Yago J. de A. Nogueira, Camila G. Mendes, Jonas J. de Paula, Débora M. Miranda, and Marco A. R.-Silva. "The Associations Between Screen Time and Mental Health in Adults: A Systematic Review." *Journal of Technology in Behavioral Science* 9 (February 17, 2024): 825–845.

Glossary

AI (Artificial Intelligence)

A broad category describing machines designed to simulate human intelligence. It includes technologies like Large Language Models (LLMs), agentic AI systems, machine learning, and more. These systems can interpret data, learn from interactions, and perform tasks that traditionally required human input.

Agentic AI

AI systems that operate with a degree of autonomy, capable of initiating actions, making decisions, and pursuing goals based on their programming or environmental feedback.

Algorithm

A set of rules or instructions given to an AI or computer system to help it solve problems or perform tasks.

Automation

The use of technology to perform tasks without human intervention, often used to reduce costs or increase efficiency, sometimes at the expense of human jobs.

Cybersecurity Drills

Simulated scenarios designed to test an organization's readiness for cyberattacks, often focusing on employee responses to phishing and social engineering.

Deepfake

Digitally altered media (images, audio, or video) created using AI that can convincingly mimic real people, often used to mislead or impersonate.

Human Attack Surface

In cybersecurity, the vulnerabilities introduced by human behaviors, emotions, and decisions that can be exploited by attackers.

Human-Centered Design

An approach to system and product development that prioritizes human needs, values, and dignity over technical optimization.

Large Language Model (LLM)

A type of AI trained on vast amounts of text data to understand and generate human-like language.

Machine Learning

A subset of AI that enables systems to learn from data and improve over time without being explicitly programmed for every outcome.

Managed Service Provider (MSP)

A company that manages a customer's IT infrastructure and end-user systems.

Prompt Injection

A security vulnerability in AI systems where input is crafted to manipulate the AI's behavior or output.

A Note of Gratitude for the Cover Art

This book's cover was not generated by artificial intelligence.

It was crafted — thoughtfully, skillfully, and entirely by hand — by Olivia, a cover artist whose work radiates intention and care.

I asked Olivia to take on the challenge of designing this cover without relying on any AI tools. She met that challenge with integrity, heart, and remarkable craftsmanship.

Her dedication to capturing the humanity within these pages is evident in every artistic detail. It's a quiet collaboration — her visual voice guiding you before mine begins.

In an era where automation is easy, Olivia reminded me that the human touch still matters. Her work stands as a beautiful testament to that truth.

If you'd like to see Olivia's portfolio, visit:

https://www.oliviamccauley.design/

About The Author

Joshua A. Gideon is a cybersecurity executive, technologist, and author who has spent nearly three decades working at the intersection of digital systems, physical security, and human-centered leadership. His career spans enterprise cybersecurity, executive protection, and teaching personal defense — experiences that have shaped his unique perspectives on risk, ethics, and the people behind the systems.

Joshua is also the co-author of *Praying Safe: The Professional Approach to Protecting Faith Communities* (with Grant Cunningham), a respected guide for safety teams in houses of worship. His work in executive protection and field training informs both the tactical and human elements of that book.

As founder of House of Gideon Publishing, Joshua supports thought-provoking, purpose-driven projects that challenge conventional thinking in technology, leadership, and personal growth. He continues to write, speak, and mentor with a focus on ethical systems design, resilient leadership, and remembering the humans often lost in the data.

He lives and works in Indiana, surrounded by a close-knit family.

Other Books By This Author

Praying Safe:
The professional approach to protecting faith communities

By Grant Cunningham and Joshua Gideon

ISBN-10: 1947404075   **ISBN-13:** 978-1947404076

Attacks against religious institutions are increasing, leaving faith communities fearful and uncertain. *Praying Safe* offers a professional, actionable guide to securing houses of worship — without resorting to hype or fearmongering. With clear, step-by-step advice, the book helps congregations assess risks, identify vulnerabilities, and implement sustainable security practices tailored to their needs. It also tackles the biggest challenge of all: how to get buy-in from skeptical members and maintain a resilient security posture over time. Written with compassion and clarity, *Praying Safe* is a must-read for any community seeking to become a hard target while preserving its spirit of welcome.

Available online or wherever books are sold.